Mussels & Clams

Mussels & Clams

PREPARE AND ENJOY DELICIOUS MEALS USING MUSSELS AND CLAMS

Anna Luccetti

NEW HOLLAND

Contents

INTRODUCTION

Let us dive deep into the magical world of the mussel, a seafood that is closely related to oysters and scallops, but with a wonderful taste of its own.

Mussels are part of a diverse group of molluscs which share the anatomical feature of a shell. While sometimes discarded in the creation of chowders, the shells of mussels have become an essential element in other recipes. They sustain the appearance of bite-size morsels. To many chefs, the shell is just as important as the flesh in the presentation of a unique culinary experience.

In their natural state, the shell of the mussel is sealed tight to guard against predators as well as the drying air during low tide. While the shells are commonly used in button making and the secreted pearls (usually of poorer quality than oysters) are found in inexpensive jewellery, it is the delicious visceral mass of the mussel that has intrigued seafood lovers worldwide.

Most species of mussels are edible, which makes them accessible to the masses, rather than remaining the exclusive domain of the affluent feasting on an exclusive catch. Mussels are versatile, and residing in the supermarket alongside tinned sardines, allow country folk even deep inland to enjoy the same tastes as their seaside counterparts. Mussels are for all to enjoy, and many have had the 'privilege' of doing just that!

Magnificent Mussels!

Blended with exotic flavors, this magnificent mollusc can form the base of soups, salads, entrees or mains. A quick browse through this book will convince you of their versatility. Seasoned with the right combination of spices, the mussel can take on a brand new hue as well as an appetizing new taste. The existence of mussels is common, but their application is broad.

This carefully selected collection of recipes is sure to cultivate a new appreciation for the mussel in even the most rigid diner. We have provided step-by-step instructions and beautiful illustrations to show you how easy it can be to impress your guests the next time you entertain. Remember that the presence of seafood on any table, instantly lifts the standard of your meal. In particular, the mussel's neat, symmetrical shell and delectable taste embellish a dinner party with an upmarket ambience. Mussels simply look good and taste great. They are an easy choice and a guaranteed winner.

If this pleasure to the palate isn't enough to convince the uncertain, then the fact that the mussel has less fat that the average T-bone steak but is still full of nutrients will entice them. With so many reasons to make the mussel an addition to your list of food favorites, come out of your shell and discover a delectable delight from the deep blue.

HARVESTING

European mussels have been cultured since the thirteenth century. Currently, a number of species of mussels are farmed globally, the most common of which is the blue mussel. China is now the largest producer of blue mussels in the world where its culture technique depends on a high proportion of spat being produced from hatcheries. Other important producers of mussels are Spain, the Netherlands, Denmark and France.

Spat Collection

Mussel spat are generally collected from wild populations by 'spat-collectors' although artificial hatchery production is technically possible. Due to the relatively low value of mussels and the ready availability of wild spat, hatchery production is generally not necessary or commercially viable.

Spat collectors are typically made of Christmas tree rope; a fibrous, 'hairy' looking rope, which is hung from long-lines in areas which are known to have good mussel 'spat-fall'. Spat collection ropes are hung in the water just in time for the peak spawning and spat-fall period. For commercial purposes, spat densities must be 500m/640ft of collection rope or greater to be economical. A spat density of 1000m/3280ft of collection rope allows for sufficient growth without overcrowding. High settlement densities of spat result in stunted growth and may result in excessive mortalities.

Juveniles are generally stripped from the ropes during January and February, when they are 12mm/½in in length, for on-growing.

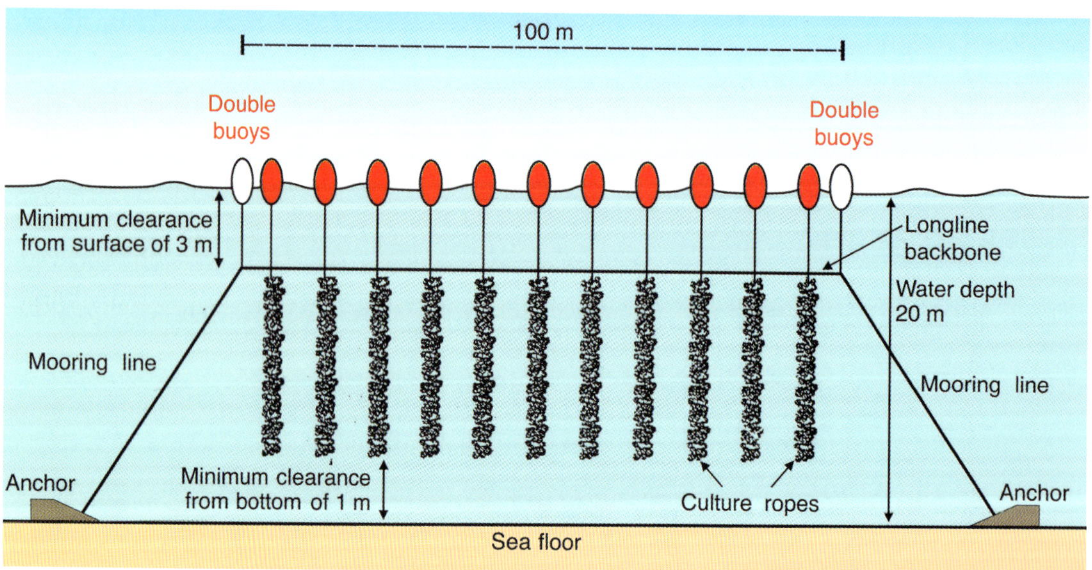

Grow-Out

Once the juveniles have reached 12mm/½in in size on the spat collectors, they are ready to be re-seeded onto grow-out longlines. Re-seeding is a process whereby the juveniles are 'thinned-out' to encourage further growth of the mussels. This is a full-time job for the farmers who, each day, weather permitting, take the collection ropes from the water to strip the mussels from Christmas tree collection rope.

The juvenile mussels are separated from each other by passing them through a mussel de-clumping machine before they are fed through a funnel and onto a grow-out rope. A cotton stocking, also known as a 'mussock' is placed around the outside of the grow-out rope holding the juvenile mussels against the rope. As the mussels grow, they re-attach themselves to the ropes by way of byssus threads. The mussock disintegrates leaving the mussels. The mussels are then hung back in the water, and continue to grow for a further 8 to 12 months. The commercial stocking rate for mussels on grow-out lines is generally between 200 and 400 mussels per metre/3½ft of rope.

It is important to reseed mussels when they are still juvenile spat because when mussels get larger than 40mm/1½in, they do not attach as readily to the rope. This means that slippage will occur with mussels slipping down and forming clumps in the bottom of the sock. This results in a reduced growth rate and poor shell shape. The

stockings and rope are skewered and tied at 0.5m/20in intervals to ensure that the mussels are evenly distributed.

Ideally, grow-out farms are situated away from heavy spat settlement areas to avoid layers of spat attaching to larger mussels. Long lines require culling to remove fouling and smothering by naturally settling juvenile mussels.

Mortalities of mussels can occur if there is a lack of food in the area or if they are exposed to extreme wave action.

Growth

The growth rate of mussels on long lines will vary according to density, depth and food availability with decreased growth experienced at high-stocking densities and greater depths.

Growth rates of blue mussels vary but can be rapid. Blue mussels can reach 32–92mm in length after 12 months and 53–110mm after 18 months. Commercial harvest usually occurs after 1 to 2 years. As in other shellfish, the meat condition changes seasonally during the growing period. Male and female blue mussels mature within 2 years of age at 4.5–5cm/1½–2in in length, with mortality at its highest during the free-floating larval stage of the life cycle.

Harvesting

Harvesting of blue mussels for market requires that the product be removed from the long lines and the shells cleaned of external fouling before presentation. This process is usually automated and involves a washer-tumbler machine in which the mussels are rotated and rubbed against each other to dislodge small mussels, barnacles and other fouling organisms. In some ideal growing areas where there are high levels of nutrients and phytoplankton, the mussels need to be transferred to more oceanic water to clean themselves before harvesting.

There is a small window of opportunity for harvest each year. Harvest must be performed once the mussels are in their best meat condition but before they spawn. It is important that prospective mussel farmers determine when they can sell their stock as this will have considerable impact on their ability to supply markets. Some areas may have a very restricted harvest period and the condition of mussels will vary annually with some years being better than others. For this reason, it is important to keep records of conditions to determine site characteristics.

Nutritional Value of Mussles (compared to T-Bone Steak)		
3.5oz/105g Cooked Meat	Common Mussels	T-Bone Steak
Calories	172	214
Protein	23.80 g	26.13 g
Fat	4.48 g	10.37 g
Carbohydrates	7.39 g	0.00 g
Cholesterol	56.00 g	80.00 g
Calcium	33.00 mg	7.00 mg
Magnesium	37.00 mg	26.00 mg
Phosphorus	285.00 mg	208.00 mg
Potassium	268.00 mg	407.00 mg
Iron	6.72 mg	3.00 mg
Omega 3 Fatty Acids	782.00 mg	0.00 mg
Omega 6 Fatty Acids	36.00 mg	290.00 mg

Soups & Salads

San Franciscan Seafood Chowder

SERVES 8

INGREDIENTS

8 smallish round loaves of bread
55 g/2 oz butter
2 leeks, finely sliced
2 onions, finely chopped
4 cloves garlic, minced
2 carrots, peeled and chopped
1 parsnip, peeled and chopped
2 ribs celery, finely sliced
1 tablespoon fresh thyme leaves
½ cup/55 g/2 oz plain flour
8 cups fish stock
1 kg/35 oz mixed seafood (including prawns (shrimp), mussels, clams, calamari (squid), white fish
200 ml/7 fl oz thickened cream
1 cup/30 g/1 oz fresh parsley, chopped
salt and pepper, to taste
juice of 1 large lemon
½ bunch chives, chopped, for garnish

METHOD

- Preheat the oven to 200°C/400°F.
- First, prepare bread for bowls. Using a sharp knife, cut a large hole in the top of the bread loaf, then remove this crusty top and set aside. Carefully remove all the soft bread from the inside of the loaf (leaving the surrounding crust intact).
- Place the loaves in the preheated oven and bake for 15 minutes (until the loaves are crisp and dry). Set aside.
- Melt the butter in a large saucepan and add the leeks, onions, garlic, carrots, parsnip, celery and thyme leaves. Sauté for 10 minutes until the vegetables are soft and golden. Remove the pan from the heat and sprinkle the flour over the vegetables, stirring constantly to mix the flour with the butter. Return the pan to the heat and continue stirring until the mixture begins to turn golden (about 2 minutes). This gives the flour a 'cooked' flavor.
- Add the fish stock stirring constantly to dissolve the roux mixture into the liquid, then simmer the soup for 20 minutes. Meanwhile, prepare the seafood by cutting the fish and shellfish into bite-sized pieces.
- Add all the seafood, cream, parsley and salt and pepper, and cook for a further 5 minutes. (Do not allow the soup to boil rapidly because it may curdle.) Once the seafood has cooked, stir the lemon juice through the fish and ladle the soup into the bread bowls. Garnish with some chopped chives and serve.

Hot and Sour Soup

SERVES 6

INGREDIENTS

4 red or golden shallots, sliced
2 fresh green chillies, chopped
6 kaffir lime leaves
4 slices fresh ginger
8 cups fish, chicken or vegetable stock
255 g/9 oz boneless firm fish fillets, cut into chunks
12 medium uncooked prawns (shrimp), shelled and deveined
12 mussels, scrubbed and beards removed
115 g/4 oz oyster or straw mushrooms
3 tablespoons lime juice
2 tablespoons Thai fish sauce (nam pla)
fresh cilantro (coriander) leaves
lime wedges

METHOD

- Place the shallots, chillies, lime leaves, ginger and stock in a saucepan and bring to the boil over a high heat. Reduce the heat and simmer for 3 minutes.
- Add the fish, prawns (shrimp), mussels and mushrooms and cook for 3–5 minutes or until the fish and seafood are cooked. Discard any mussels that do not open after 5 minutes of cooking. Stir in the lime juice and fish sauce. To serve, ladle the soup into bowls, scatter with cilantro (coriander) leaves and accompany with lime wedges.

Clam Chowder

SERVES 6–8

INGREDIENTS

255 g/9 oz butter
6 rashes of bacon, finely chopped
3 onions, finely chopped
1½ cups finely chopped celery
¼ cup plain flour
4 cups milk
3 cups fish stock
500 g/1 lb potatoes, finely diced
1 kg/35 oz clams
salt and pepper
fresh parsley, finely chopped, to garnish
1 French loaf, cut into cubes, lightly fried

METHOD

- Heat the butter in a saucepan and cook the bacon, onion and celery until tender.
- Add the flour and cook for 2 minutes.
- Add the milk, fish stock and potatoes, cover and simmer for 10 minutes.
- Add the clams and cook again for 10 minutes. Season to taste.
- Serve in a deep plate with a sprinkle of parsley and croutons.

Mussel Soup

SERVES 4

INGREDIENTS

olive oil, for frying
½ onion, finely diced
55 g/2 oz butter
75 ml/2½ fl oz milk
100 ml/3½ fl oz dry white wine
2 tablespoons plain flour
1 kg/35 oz mussels, cooked mariniéres style (see page 100)
chives, to garnish

METHOD

- Add oil to a frypan over moderate heat. Add onion and gently fry until soft.
- In a cooking pot over medium heat, melt the butter then add the flour. Stir with a wooden spoon and cook gently for 2 minutes.
- Add the milk slowly with a whisk and cook until slightly thickened and smooth in consistency.
- Add the wine and mussels and bring to the boil. Add salt and pepper to taste. Garnish with chives just before serving.

Clam Bisque

SERVES 4–6

INGREDIENTS

455 g/1 lb white fish fillets
3 cups milk
salt and pepper, to taste
⅛ teaspoon nutmeg
1 bay leaf
250 g/8 oz jar mussels
2 tablespoons butter
1 medium-sized onion, finely chopped
2 sticks celery, finely cubed
3 tablespoons plain flour
1 tablespoon lemon juice
1 tablespoon finely chopped parsley or chives
1–2 tablespoons dry sherry
¼ cup cream

METHOD

- Cut the fish fillets into 2 cm/¾ in squares. Place in a saucepan with the milk, salt, pepper, nutmeg and bay leaf. Bring gently to the boil, then simmer slowly for 10 minutes. Stand covered for 10 minutes to infuse the flavors. Strain the milk from the fish and reserve. Keep the fish warm.
- Drain the liquid from the mussels and rinse in cold water. Cut the mussels into 2 or 3 pieces.
- Melt the butter in a large saucepan, add the onion and celery and cook gently for 10 minutes without browning. When soft, stir in the flour and cook for 1 minute while stirring.
- Remove the saucepan from the heat and gradually stir in the reserved milk, stirring well after each addition until free from lumps. Return to the heat and stir until the mixture boils and thickens.
- Add the lemon juice, chopped mussels, chopped parsley, sherry and cooked fish. Simmer slowly for 10 minutes. Stir in the cream and simmer for 5 minutes more. Serve in individual bowls, with croutons if desired.

Clam and Black Mussel Broth

SERVES 6

INGREDIENTS

45 ml/1½ fl oz vegetable oil
1 onion finely chopped
2 tablespoon tom yum paste
200 g/7 oz surf clams, cleaned and free of sand
200 g/7 oz black mussels, cleaned
1 cup chicken stock
1 stalk lemongrass, chopped
juice of 1 lime
1 tablespoon cilantro (coriander) stalk and roots, finely chopped
1 tablespoon Thai fish sauce
1 tablespoon fresh cilantro (coriander) leaves, roughly chopped

METHOD

- Heat the oil in a wok or large cooking pot. Add the onion, tom yum, clams and mussels. Simmer, covered with a lid, for 30 seconds.
- Add the chicken stock, lemongrass, lime juice, cilantro (coriander) stalk and roots, Thai fish sauce and stir through. Cook until all shells have opened.
- Add the fresh cilantro (coriander) leaves and serve in soup bowls.

Mussels, Witlof and Basil Soup

SERVES 4

INGREDIENTS

25 g/1 oz butter
2 witlof or Belgian endives, leaves cut
1 cup broth from cooking the mussels, strained
2 tablespoons double cream or thickened cream
15 basil leaves, chopped finely
salt and pepper, to taste
1 kg/2¼ lb black mussels, cooked mariniéres style and removed from shell (see page 100)

METHOD

- Melt the butter in a pot and stir-fry the witlof.
- Add the broth, cream and basil. Mix well and whip until the ingredients are all blended. Season with salt and pepper.
- Add the mussels, heat until boiling and serve in bowls.

Maltese Mussels

SERVES 6

INGREDIENTS

100 ml/3⅓ fl oz olive oil
1 onion, diced
3 garlic cloves, chopped
2 roasted bell peppers (capsicums), peeled and diced
½ bulb fennel, diced
½ celery stick, diced
145 ml/5 fl oz white wine
1 orange juice plus grated zest
200 ml/7 fl oz tomato juice
2 cups chicken stock
salt and pepper
cayenne pepper and paprika
455 g/1 lb black mussels, cleaned

METHOD

- In a large pot on medium heat, add the oil, onion, garlic, bell pepper (capsicum), fennel and celery. Cook for 10 minutes.
- Add the white wine, orange juice, zest, tomato juice, chicken stock, seasoning and cook for 20 minutes.
- Add the mussels and cook until all the mussels have opened, about 8–10 minutes.
- Serve with wood-fired bread or grissini.

Manhattan Chowder

SERVES 4–6

INGREDIENTS

2 kg/4½ lb mussels
4 rashers bacon, rind removed and diced
1 large onion, chopped
1 bay leaf
1 green bell pepper (capsicum), finely diced
2 stalks celery, diced
455 g/1 lb peeled and diced potatoes
2 x 400 g/14 oz cans peeled tomatoes, seeds removed, chopped
Tabasco sauce
salt and pepper to taste

METHOD

- Wash and scrub the mussels and steam them open. Remove and discard the shells, reserving the mussels. Strain the cooking liquor through fine muslin and add enough water to measure 4 cups.
- In a heavy pan, cook the bacon gently until it begins to crisp. Add the onion and the bay leaf and sauté until the onion is tender, about 8 minutes. Add the bell pepper (capsicum) and celery and sauté for a further few minutes. Add the potatoes and tomatoes with the juice, the mussel cooking liquid, and salt and pepper to taste. Bring to the boil and simmer, covered, for 20 minutes, until the potatoes are tender. Discard the bay leaf, add the mussels and cook a further 5–10 minutes. Season the chowder with a little Tabasco.

Cold Marinated Mussel Salad

SERVES 4–6

INGREDIENTS

1 small carrot, diced finely
55 g/2 oz cauliflower, broken into florets
½ red bell pepper (capsicum), diced finely
½ onion, diced finely
pinch saffron
10 cilantro (coriander) seeds, cracked
45 ml/1½ fl oz sherry vinegar
285 g/10 oz cooked mussel meat (equivalent to approximately 1 kg/2 lb mussels in their shells, cooked mariniéres style and chilled (see page 100)
285 ml/10 fl oz water

Salad

handful mesclun salad mix
cherry tomatoes, quartered
3 tablespoons virgin olive oil
salt and pepper

METHOD

- Place the water, carrot, cauliflower, bell pepper (capsicum), onion, saffron and cilantro (coriander) seeds in a pot over high heat.
- Bring to the boil and add the vinegar.
- Remove from the heat straight away and allow to cool down. When cold, strain the vegetables and discard the cooking liquid.
- In a large salad bowl, mix together the salad, tomatoes, olive oil, vegetables and mussels. Season with salt and pepper.

Scandinavian Mussels

SERVES 2

INGREDIENTS

100 ml/3½ fl oz water
½ onion, finely chopped
½ stick celery, finely chopped
½ red bell pepper (capsicum), finely chopped
1 tablespoon sugar
2 tablespoon white vinegar
1 kg/2¼ lb black mussels cleaned, cooked mariniéres style and removed from the shell (see page 100)
4 tablespoons mayonnaise
1 tablespoon chopped parsley
juice of 1 lemon
salt and pepper to taste

METHOD

- In a small cooking pot, bring water, onion, celery, bell pepper (capsicum), sugar and vinegar to the boil for 1 minute and set aside to cool.
- Remove vegetables from liquid.
- Mix mussels, mayonnaise, vegetables, parsley and lemon juice in a bowl, add salt and pepper.
- Serve cold with green salad or a cold potato salad.

Seafood Paella Salad

SERVES 6

INGREDIENTS

4 cups chicken stock
510 g/18 oz uncooked large prawns (shrimp)
1 uncooked lobster tail (optional)
510 g/18 oz mussels in shells, cleaned
2 tablespoons olive oil
1 onion, chopped
2 ham steaks, cut into 1cm/½ in cubes
2 cups/400 g/14 oz arborio rice
½ teaspoon ground turmeric
115 g/4 oz fresh or frozen peas
1 red bell pepper (capsicum), diced

Garlic Dressing

½ cup olive oil
¼ cup white wine vinegar
3 tablespoons mayonnaise
2 cloves garlic, crushed
2 tablespoons chopped fresh parsley
freshly ground black pepper

METHOD

- Place the stock in a large saucepan and bring to the boil. Add the prawns (shrimp) and cook for 1–2 minutes or until the prawns change color. Remove and set aside. Add the lobster tail and cook for 5 minutes or until the lobster changes color and is cooked. Remove and set aside. Add the mussels and cook until the shells open – discard any mussels that do not open after 5 minutes. Remove and set aside. Strain the stock and reserve. Peel and devein the prawns (shrimp), leaving the tails intact. Refrigerate the seafood until just prior to serving.
- Heat the oil in a large saucepan, add the onion and cook for 4–5 minutes or until soft. Add the ham, rice and turmeric and cook, stirring, for 2 minutes. Add the reserved stock and bring to the boil. Reduce the heat, cover and simmer for 15 minutes or until the liquid is absorbed and the rice is cooked and dry. Stir in the peas and red bell pepper (capsicum) and set aside to cool. Cover and refrigerate for at least 2 hours.
- To make the dressing, place the oil, vinegar, mayonnaise, garlic, parsley and black pepper to taste in a food processor or blender and process to combine.
- To serve, place the seafood and rice in a large salad bowl, spoon over the dressing and toss to combine.

Mussel and Rice Salad

SERVES 6–8

INGREDIENTS

3 cups chicken stock
1½ cups/330 g/11 oz long-grain rice
2 kg/4½ lb mussels, scrubbed
¾ cup/180 ml/6 fl oz white wine
2 teaspoons chopped tarragon or 1 teaspoon dried

Dressing

1 tablespoon wine vinegar
2 teaspoons Dijon mustard
salt and freshly ground pepper
4 tablespoons olive oil
2 tablespoons chopped parsley

METHOD

- Bring the stock to the boil and add in the rice. Bring to the boil again, cover and cook gently until the rice is tender and the stock absorbed, about 18 minutes. Steam the mussels in the wine, tarragon and a little water. Remove the mussels from their shells leaving a few in the half shell for garnishing, if desired.
- Add the mussels to the rice. Pour off and discard half the cooking liquid and strain remainder into a pan. Reduce over a high heat and sprinkle over the rice and mussels.
- Mix the vinegar and mustard together with salt and pepper. Whisk in the oil, a little at a time, to make a thick dressing. Add to the salad and toss lightly. Cool to room temperature and toss gently with the parsley.
- Arrange on 6 serving plates with salad greens.

Mussels and Prawns in Avocado Vinaigrette

SERVES 2

INGREDIENTS

1 dozen medium mussels, cooked mariniéres style, reserving some of the broth (see page 100)
8 large green prawns (shrimp), shells removed
shredded lettuce
minced parsley

Dressing

1 avocado
2½ tablespoons fresh lemon juice
4 tablespoons olive oil
salt
freshly ground pepper
¼ teaspoon Dijon-style mustard
1 teaspoon grated onion
¼ teaspoon paprika

METHOD

- Shell the mussels and to keep them moist, place them in a bowl with a little of the liquid in which they have cooked. Cook the prawns in the remaining mussel liquid for 1–2 minutes. Cool and shell.
- To prepare the dressing, peel the avocado and remove the pit. Cut the avocado in pieces and place in a processor with the lemon juice. Blend until smooth. Gradually add the olive oil. Season very well with salt and pepper and add the mustard, onion and paprika. (May be prepared ahead.)
- Arrange the mussels and prawns attractively on a bed of shredded lettuce. Cover with the avocado dressing and sprinkle with parsley. Serve cold or at room temperature.

Witlof Speck Salad with Clams and Mussels

SERVES 6–8

INGREDIENTS

400 g/14 oz clams (pipies) cooked and out of the shells
400 g/14 oz mussels cooked and out of the shells
2 witlof, cut with leaves loose
4 slices of prosciutto, cooked under the grill and broken into small pieces
45 ml/1½ fl oz virgin olive oil
1 tablespoon lemon juice
salt and pepper, to taste

Marinated Vegetables

1 carrot, peeled and sliced
½ onion, sliced in half
½ stick celery
15 cilantro (coriander) seeds, cracked
salt and pepper
30 ml/1 fl oz sherry vinegar
1 cup/250 ml/8 fl oz water

METHOD

- To prepare the marinated vegetables, place the ingredients together in a pot and boil for 2 minutes. Set aside to cool. When cold remove the liquid and keep the vegetables aside.
- Mix all the other ingredients together with the marinated vegetables.
- Refrigerate for 15 minutes until chilled.
- Serve with crispy bread or Italian Grissini.

Mixed Shellfish and Potato Salad

SERVES 4

INGREDIENTS

680 g/1½ lb waxy potatoes, unpeeled
4 small cooked beetroot, diced
1 head fennel, finely sliced, plus feathery top, chopped
1 kg/2¼ lb mussels
510 g/18 oz clams
285 ml/10 fl oz dry white wine 1 shallot, finely chopped
4 spring onions (scallions), finely sliced
3 tablespoons chopped fresh parsley

Dressing

5 tablespoons olive oil
1 tablespoons cider vinegar
1 teaspoon English mustard

METHOD

- To make the dressing, whisk together the oil, vinegar, mustard and seasoning. Boil the potatoes in salted water for 15 minutes or until tender, then drain. Cool for 30 minutes, then peel and slice. Place in a bowl and toss with half the dressing. Toss the beetroot and fennel with the rest of the dressing.
- Scrub the mussels and clams under cold running water, pulling away any beards from the mussels. Discard any shellfish that are open or damaged. Place the wine or cider and shallot in a large saucepan and bring to the boil. Simmer for 2 minutes, then add the shellfish. Cover and cook briskly for 3–5 minutes, shaking the pan often, or until the shellfish have opened. Discard any that remain closed. Reserve the pan juices, set aside a few mussels in their shells and shell the rest.
- Boil the pan juices for 5 minutes or until reduced to 1–2 tablespoons. Strain over the potatoes. Add the shellfish, spring onions (scallions) and parsley, then toss. Serve with the beetroot and fennel salad and garnish with the fennel tops and mussels in their shells.

Light Meals

Clam Provençal

SERVES 4–6

INGREDIENTS

100 ml/3⅓ fl oz virgin olive oil
1 onion, finely chopped
1 bell pepper (capsicum), diced
4 vine-ripened tomatoes, diced
½ stick celery, sliced
2 cloves garlic
1 kg/35 oz clams, cleaned and sand removed
145 ml/5 fl oz dry white wine
1 tablespoon freshly chopped aromatic herbs (thyme, rosemary, marjoram)
salt and pepper

METHOD

- Place the oil, onion, bell pepper (capsicum), tomatoes, celery and garlic in a large cooking pot. Cook over high heat for 5 minutes, stirring frequently to prevent sticking.
- Add the clams, white wine, fresh herbs and seasoning and cook with the lid on until all the shells have opened. Stir frequently to ensure even cooking.
- When the clams are open, serve in large bowls with salad or grilled, sliced baguette. Complement the dish with a rosé or white wine from Provence.

Smoked Mussel Fritters

SERVES 4–6

INGREDIENTS

2 x 85 g/3½ oz cans smoked mussels, drained
125 g/4½ oz plain flour
1 teaspoon baking powder
1 teaspoon salt
1 small red bell pepper (capsicum), seeded and finely diced
2 tablespoons chopped fresh cilantro (coriander)
½ teaspoon cayenne pepper
1 egg, lightly beaten
55 ml/2 fl oz beer
1 tablespoon lime juice
vegetable oil, for deep frying
lime wedges, to serve

Sauce

4 tablespoons mayonnaise
1 teaspoon wholegrain mustard
juice ½ lime
½ teaspoon clear honey

METHOD

- Roughly chop the mussels and place in a bowl with the flour, baking powder, salt, red bell pepper (capsicum), cilantro (coriander) and cayenne and stir well.
- Beat in the egg, beer and lime juice to form a soft dropping batter.
- Heat the oil in a wok or deep frying pan over high heat. Drop tablespoonfuls of the mussel batter in batches into the hot oil. Fry for 3 minutes until golden. Drain on kitchen paper and keep warm in a moderate oven, while cooking the rest.
- Mix all the sauce ingredients together in a bowl. Serve the fritters with the sauce and some lime wedges to squeeze over.

Grilled Mussels

SERVES 4

INGREDIENTS

20 large fresh mussels
30 g/1 oz basil leaves
1 clove garlic, crushed
1 small red chilli, de-seeded and diced ½ teaspoon grated lemon rind
1 tablespoon pine nuts
1 tablespoon Parmesan cheese
1 tablespoon fresh breadcrumbs
salt and pepper
3–4 tablespoons extra virgin olive oil

METHOD

- Wash and scrub the mussels. Steam the mussels with just the water on their shells, for 4 minutes until they have just opened. Discard any which do not open. Immediately plunge the mussels into cold water and drain again.
- Remove the mussels from the pan and carefully discard one half of each shell. Arrange the remaining mussels in 1 large dish or 4 individual gratin dishes.
- In a blender or food processor combine the basil, garlic, chilli, lemon rind, pine nuts, Parmesan and half of the breadcrumbs. Pulse briefly to form a smooth paste and season to taste with salt and pepper.
- Transfer the basil paste to a bowl and stir in the oil. Spoon a little of the paste over each mussel and finally top each one with a few more breadcrumbs. Cook under a preheated grill for 2–3 minutes until bubbling and golden. Serve at once.

Baked Mussels

SERVES 4

INGREDIENTS

30 medium mussels
2 shallots, finely chopped
1 sprig thyme
2 sprigs parsley
1 bay leaf
½ teaspoon salt
½ cup white wine
115 g/4 oz butter or margarine, softened
1 tablespoon parsley, chopped
2 cloves garlic, crushed
1 tablespoon chives

METHOD

- Scrape the beard, and wash the mussels thoroughly and place in a large saucepan with the shallots, thyme, parsley and bay leaf.
- Sprinkle the salt over and then add the wine. Steam for 5 minutes or until the shells have opened. Open the mussels and discard the lids.
- Divide the mussels in the remaining half shells into 4 ovenproof dishes. Make a herb butter by combining the butter, parsley, garlic and chives and place a generous portion on each mussel.
- Bake at 190°C/370°F for approximately 3 minutes or until the butter has melted.

Black Forest Mussels with Mushrooms and Brandy

SERVES 4

INGREDIENTS

30 g/1 oz butter
½ onion, finely chopped
145 g/5 oz finely sliced black forest mushrooms or field mushrooms
1 clove garlic, chopped
1 kg/2¼ lb black mussels, cleaned
100 ml/3½ fl oz white wine
salt and pepper to taste
2 tablespoons thickened cream or double cream
30 ml/1 fl oz brandy

METHOD

- Place the butter, onions, mushrooms and garlic in a pot and cook over high heat for 5 minutes.
- Add the mussels, white wine and seasoning.
- Cook the mussels until all mussels have opened, stirring frequently.
- Add the cream and stir for 30 seconds. Add the brandy and cook for another 1 minute.
- Serve garnished with fresh parsley if desired.

Fried Vongole

SERVES 4

INGREDIENTS

2 eggs, lightly beaten
salt and pepper
2 cups/225 g/8 oz breadcrumbs
1 tablespoon dry mixed aromatic herbs
1 kg/2¼ lb clams cleaned, cooked mariniéres style and removed from shell (see page 100)
3 tablespoons tartare sauce

METHOD

- Place the eggs in a bowl and season with a little salt and pepper.
- Combine the breadcrumbs and herbs in a separate bowl.
- Dip the clams in the egg mixture, then roll in breadcrumbs.
- Deep-fry the clams in hot oil, until golden brown.
- Drain on absorbent paper and serve immediately with tartare sauce.

Clams with White Wine and Garlic

SERVES 4

INGREDIENTS

680 g/1½ lb medium to small clams
salt and pepper
45 ml/1½ fl oz olive oil
1 small onion, peeled and finely chopped
3 cloves garlic, peeled and minced
1 tablespoon plain flour
100 ml/3½ fl oz dry white wine
pinch of paprika
1 bay leaf

METHOD

- Wash the clams well. Leave them in cold water to which you have added a little salt for about an hour to get rid of any grit they may have in them.
- Heat the oil in a frying pan. Add the onion and garlic and sauté until golden brown. Add the clams and cook over medium heat until the shells open. Add the flour and stir in well. Pour in the wine, add the paprika, bay leaf and some salt and pepper to taste. Continue cooking for a further 5 minutes.
- Remove the bay leaf and serve the clams in the sauce. Half the fun is to eat the clam and then spoon up some sauce using the shell as a spoon.

Seafood Escabêche

SERVES 4

INGREDIENTS

16 large mussels, scrubbed
225 g/8 oz baby calamari (squid), cleaned
225 g/8 oz peeled small prawns (shrimp), deveined
2 cloves garlic, chopped
1 small red chilli, de-seeded and chopped
55 ml/2 fl oz dry sherry
1 tablespoon chopped basil, to garnish
bread, to serve

Marinade

145 ml/5 fl oz extra virgin olive oil
2 shallots, chopped
3 tablespoons white wine vinegar
pinch sugar
1 tablespoon drained and chopped capers in brine
salt and pepper

METHOD

- Prepare the seafood. Remove any beard still attached to the mussels. Cut the squid into rings and the tentacles in half, if large and wash well. Wash and dry the raw prawns (shrimp).
- Place the mussels into a pan with the garlic, chilli and sherry. Cover and steam for 4–5 minutes until all the mussels are opened (discard any that remain closed). Remove the mussels with a slotted spoon and set aside.
- Poach the prawns (shrimp) in the mussel liquid for 4–5 minutes until cooked. Poach the calamari (squid) for 2–3 minutes until cooked. Remove with a slotted spoon and add to the mussels. Reserve 2 tablespoons of the cooking liquid and leave to cool.
- Combine all the marinade ingredients and stir in the reserved poaching liquid. Pour over the cold seafood, toss well and chill for several hours.
- Return the escabêche to room temperature for 1 hour, scatter over the basil and serve with bread.

Mussels with Tomatoes and Wine

SERVES 4

INGREDIENTS

1 kg/2¼ lb fresh mussels, scrubbed and beards removed
1 shallot, chopped
1 cup dry white wine
chopped fresh chives

Tomato and Smoked Salmon Sauce

2 teaspoons olive oil
2 cloves garlic, crushed
2 shallots, chopped
2–3 slices smoked salmon, sliced into thin strips
1 red bell pepper (capsicum), sliced
1 tablespoon no-added-salt tomato paste
400 g/14 oz canned no-added-salt diced tomatoes
2 tablespoons chopped fresh parsley

METHOD

- For the sauce, heat the oil in a non-stick frying pan over a medium heat. Add the garlic and shallots. Cook, stirring, for 1–2 minutes. Add the salmon and red bell pepper (capsicum). Cook, stirring, for 3 minutes. Stir in the tomato paste. Cook for 3–4 minutes or until it becomes deep red and develops a rich aroma. Add the tomatoes. Cook, stirring, for 5 minutes or until the mixture starts to thicken. Stir in the parsley. Keep warm.
- Meanwhile, place the mussels, shallot and wine in a large saucepan over a high heat. Cover. Bring to the boil then reduce the heat. Cook for 5 minutes or until the mussels open. Discard any mussels that do not open after 5 minutes of cooking.
- Add the sauce to the mussels. Toss to combine.
- To serve, divide the mixture among deep bowls. Scatter with chives. Accompany with crusty bread and a glass of red wine.

Seafood Stew

SERVES 6–8

INGREDIENTS

¼ cup olive oil
12 mussels in their shell, scrubbed
2 onions, thinly sliced
salt and pepper
3 tomatoes, sliced thickly
1 kg/2¼ lb white fish fillets
225 g/8 oz calamari (squid), cut into strips or rings
1 green bell pepper (capsicum), sliced
1 cup dry white wine
3 tablespoons chopped parsley

METHOD

- Pour a little oil into a large a heavy based saucepan and arrange the scrubbed mussels over the base. Cover with a layer of onion rings and sprinkle with a little salt and pepper.
- Place the tomato slices over the onions and season. Drizzle a little oil over the tomatoes. Place the fish and calamari (squid) in layers over the tomato. Season each layer and drizzle with a little oil.
- Place a layer of green bell pepper (capsicum) rings on top. Pour in the white wine.
- Cover and cook over a slow heat for 20–25 minutes until the fish flakes.
- Spoon into individual soup bowls and sprinkle with chopped parsley. Serve with crusty bread.

Clams with Vinegar

SERVES 4

INGREDIENTS

285 g/10 oz can baby clams, drained and well chilled

1 tablespoon white wine vinegar

METHOD

- Simply mix the clams with the vinegar and serve cold as part of an antipasto plate.

Spaghettini with Baby Clams, Chilli and Garlic

SERVES 4–6

INGREDIENTS

400 g/14 oz spaghettini
55 ml/2 fl oz olive oil
4 cloves garlic, sliced
4 red chillies, finely chopped
2 cups tomato, finely diced
680 g/24 oz canned baby clams or fresh, if available
⅓ cup/10 g/1 oz parsley, chopped
juice of 2 lemons
salt and freshly ground black pepper

METHOD

- Cook the spaghettini in boiling water with a little oil until al dente. Run under cold water until cold and set aside.
- Heat half the oil and cook the garlic on a low heat until beginning to change color. Add the chillies and tomatoes, and cook for a few minutes.
- Add the clams, parsley, lemon juice, remaining oil, spaghettini and a little of the water used to cook the clams, and heat through for a further 5 minutes. Season with salt and black ground pepper to taste.
- Note: If using fresh clams, wash under running water, scraping the shells with a sharp knife or scourer. Place them in a large pan with a little water over a gentle heat until they open. Discard any that do not open.

Leek and Seeded Mustard Mussels

SERVES 4

INGREDIENTS

30 g/1 oz butter
1 leek sliced finely (do not use all the green leaves; cut halfway as they get bitter at the end of the plant.)
1 kg/2¼ lb black mussels, cleaned
3 tablespoons seeded mustard
115 ml/4 fl oz dry white wine
2 tablespoons thickened cream or double cream
1 tablespoon chopped parsley
salt and pepper

METHOD

- Place butter in a pot over high heat. Add leeks and cook for 1 minute.
- Add mussels, seeded mustard and white wine.
- Cook mussels until all have opened, stirring frequently for evenly cooking.
- Add cream and parsley, stir for 15 seconds on high heat. Season with salt and pepper to taste.

Mussel and Zucchini Gratin

SERVES 6

INGREDIENTS

2 kg/4½ lb mussels
small bunch finely chopped shallots
bay leaf
⅔ cup dry white wine
1 kg/2¼lb zucchini (courgette)
salt and freshly ground pepper
¼ cup olive oil
1¼ cups cream
3 egg yolks
2 tablespoons grated Gruyére cheese

METHOD

- Wash and scrub the mussels well with a stiff brush. With a firm tug, pull the small piece of sea grass from the side of each mussel. Soak for 1 hour or so in water to allow the mussels to disgorge any sand.
- Place the mussels in a large pan with the shallots and bay leaf, cover and cook over a brisk heat for 5 minutes or until the shells have opened. Remove with a slotted spoon. Taste the cooking juices and, if too salty, discard half.
  Add the wine to the pan and reduce the liquid to about ½ cup. Strain and reserve. Shell the mussels, if they are large discard the black rims, and set aside.
- Wash, trim and slice the zucchini (courgette). Season with salt and pepper and sauté gently in the olive oil in a large frying pan until lightly browned. Transfer to a large gratin dish. Preheat the oven to 225°C/425°F.
- Meanwhile, reduce 1 cup of the cream over a gentle heat to ¾ cup and stir in the reserved mussel cooking liquor. In a small bowl, beat the egg yolks with the remaining cream and stir in 2 tablespoons of the hot reduced cream mixture. Stir this mixture into the cream sauce in the pan, then remove from the heat. Check the seasoning.
- Top the zucchini (courgette) with the mussels then the cream sauce. Sprinkle with the cheese and heat in the preheated oven for 10 minutes, or until the top is browned.

Mussels in Shells with Herb Mayonnaise

SERVES 4

INGREDIENTS

1½ kg/3⅓ lb small mussels
6 spring onions (scallions), chopped
2 bay leaves, crumbled
fresh basil leaves or chervil sprigs for garnish (optional)

Herbed Mayonnaise

1 large egg at room temperature
1½ tablespoons fresh lemon juice
1½ teaspoons Dijon-style mustard
1 cup/250 ml/8 fl oz olive oil
1 tablespoon tomato paste
2 cloves garlic, mashed to a paste with 1 teaspoon salt
pepper, to taste
30 g/1 oz chopped herbs: mostly parsley, some basil or oregano, a little thyme

METHOD

- Make the mayonnaise. In a blender with the motor on high or in a food processor, blend the egg, lemon juice and mustard. Add the oil in a slow stream until a thick emulsion forms. Add the tomato paste, garlic and pepper to taste and blend the mayonnaise with the herbs until it is combined well. Transfer the mayonnaise to a small bowl. Cover and chill.
- Scrub the mussels well in several changes of water, scrape off the beards and rinse the mussels. In a large pan, steam the mussels with 1 cup of water, the spring onions (scallions) and bay leaves, covered, over a moderately high heat for 5–7 minutes, or until the mussels are opened. Transfer them with a slotted spoon to a bowl, and discard any unopened mussels.
- Remove the mussels from the shells, discarding half the shells, and arrange 1 mussel in each of the remaining shells.
- Transfer the mussels to a large platter and chill them, covered, for 20 minutes. Just before serving, spoon about 1 teaspoon of the herb mayonnaise over each mussel and sprinkle the mussels with the basil, which should be finely shredded at the last moment.

Pasta Frutti Di Mare

SERVES 4

INGREDIENTS

510 g/18 oz spaghetti
2 teaspoons vegetable oil
2 teaspoons butter
½ cup dry white wine
500 g/17½ oz) clams (pipies)
12 mussels, scrubbed and beards removed
12 scallops
12 uncooked prawns (shrimp), shelled and deveined
1–2 garlic cloves, crushed
handful parsley, chopped, for cooking and garnish

METHOD

- Cook the pasta in boiling water in a large saucepan following the packet directions. Drain, set aside and keep warm.
- Heat the oil and butter in a frying pan over medium heat. Add crushed garlic and half a handful of the parsley and saute until garlic is translucent for 4 minutes or until the onions are golden.
- Add the wine and bring to the boil. Add the clams and mussels then the scallops and prawns last, cover and cook till the prawns are pink, about 5–6 minutes. Spoon pasta into serving bowls. Add equal amounts of the seafood over the hot pasta and serve immediately.

Clams in Sherry Sauce

SERVES 1

INGREDIENTS

2 tablespoons olive oil
1 onion, finely chopped
55 g/2 oz cubed cured ham
2 tablespoons semi-sweet (oloroso) Spanish sherry
1 dozen very small clams

METHOD

- Heat the oil in a small frying pan, sauté the onion for 1 minute, then cover and cook very slowly until the onion is tender but not colored (about 15 minutes). Stir in the ham, then add the sherry and the clams.
- Turn the heat up to medium, cover and cook, removing the clams as they open. Return them to the sauce, making sure the clam meat is covered by the sauce. (May be prepared ahead.)

Clams with Pasta

SERVES 2

INGREDIENTS

2 tablespoons olive oil
1 large onion, finely chopped
1 clove garlic, minced
1 red capsicum (bell pepper), finely chopped
1 medium tomato, skinned, seeded and finely chopped
1 tablespoon chopped parsley
1 bay leaf
few strands of saffron
salt
freshly ground pepper
225 g/8 oz spaghetti, broken into 3 lengths
¾ cup veal broth, or a mixture of chicken and beef broth
2 dozen very small clams
3 tablespoons fresh or frozen peas

METHOD

- Heat 1 tablespoon of the oil in a frying pan. Sauté the onion, garlic and red bell pepper (capsicum) for 1 minute, then cover and cook slowly until the vegetables are tender but not brown (about 20 minutes). Add the tomato, parsley, bay leaf, saffron, salt and pepper to the onion mixture and cook for 5 minutes, uncovered.
- Meanwhile, bring a large pot of salted water to the boil with the remaining tablespoon of oil. Add the spaghetti and cook, stirring occasionally. Drain the spaghetti and return to the pot. Combine the onion and tomato mixture in the pot with the spaghetti, add ½ cup/120ml/4fl oz of the broth, along with the clams and the peas. Mix well. Cover and continue cooking until the clams have opened (about 10 minutes).
- To serve, add the remaining broth (the mixture should be a little soupy) and season with salt and pepper – it should be well seasoned. Serve in small individual casserole dishes. Although best prepared at the last minute, this dish can be made in advance and reheated.

Mussels Mariniéres

SERVES 3–4

INGREDIENTS

1 kg/2¼ lb mussels, cleaned
1 small onion, sliced
1 stick of celery, sliced
1 clove garlic, chopped
55 ml/2 fl oz water or white wine
pepper
1 tablespoon butter
1 tablespoon parsley, chopped

METHOD

- Place the mussels, onion, celery, garlic and water (or white wine) in a large saucepan.
- Cook over medium heat until the mussels have opened. Stir frequently to ensure the mussels cook evenly.
- Add pepper to taste. Stir in the butter and parsley just before serving.

Mussels Poulette

SERVES 3–4

INGREDIENTS

30 g/1 oz butter
1 onion, chopped
115 g/4 oz mushrooms, sliced
55 g/2 oz bacon, diced
1 stick of celery, sliced
1 kg/2¼ lb mussels, cleaned
55 ml/2 fl oz white wine
200 g/7 fl oz thickened or double cream

METHOD

- Place the butter, onion, mushrooms, bacon and celery in a saucepan over medium heat. Cook for 3 minutes.
- Add the mussels and white wine. Cook until the mussels are opened, stirring frequently.
- Stir in the cream and cook for a further minute. Serve immediately.

Mussels Provençal

SERVES 2

INGREDIENTS

1 tablespoon olive oil
½ red bell pepper (capsicum), chopped
2 tomatoes, roughly chopped
1 clove garlic, chopped
1 stick of celery, sliced
1 onion, chopped
1 kg/2¼ lb mussels, cleaned
45 ml/1½ fl oz white wine
1 tablespoon chopped basil

METHOD

- Place the oil, bell pepper (capsicum), tomatoes, garlic, celery and onion in a saucepan over medium heat. Cook for 4 minutes.
- Add the mussels and white wine. Cook until the mussels are opened, stirring frequently.
- Stir in the basil just before serving.

Mussels in Vinaigrette

SERVES 6–8

INGREDIENTS

2 kg/4½ lb mussels
salt and pepper
4 sprigs of parsley, chopped

Vinaigrette Sauce

85 ml/3fl oz olive oil
45 ml/1½ fl oz wine vinegar
15 g/½ oz finely chopped onion
15 g/½ oz finely chopped canned or home prepared sweet red bell peppers (capsicums)
15 ml/½ oz finely chopped parsley

METHOD

- Scrub and scrape the mussels well and remove the beards. Throw away any mussels that are open and do not close when tapped with a knife. Wash well in water and drain.
- Place the mussels in a saucepan with 1 cup/250 ml/8 fl oz of cold water and a pinch of salt. Place over high heat, cover and bring to the boil. Remove the mussels as they open and leave to cool. Throw away any that do not open.
- Make the vinaigrette sauce by mixing the oil, vinegar, onion, sweet red bell peppers (capsicums), parsley and some salt and pepper in a bowl. Arrange the mussels in a dish and spoon the sauce over each one.

Clams with Pasta

SERVES 2

INGREDIENTS

2 tablespoons olive oil
1 large onion, finely chopped
1 clove garlic, minced
1 red bell pepper (capsicum), finely chopped
1 medium tomato, skinned, seeded and finely chopped
1 tablespoon chopped parsley
1 bay leaf
few strands of saffron
salt
freshly ground pepper
225 g/8 oz spaghetti, broken into 3 lengths
¾ cup veal broth, or a mixture of chicken and beef broth
2 dozen very small clams
3 tablespoons fresh or frozen peas

METHOD

- Heat 1 tablespoon of the oil in a frying pan. Sauté the onion, garlic and red bell pepper (capsicum) for 1 minute, then cover and cook slowly until the vegetables are tender but not brown (about 20 minutes). Add the tomato, parsley, bay leaf, saffron, salt and pepper to the onion mixture and cook for 5 minutes, uncovered.
- Meanwhile, bring a large pot of salted water to the boil with the remaining tablespoon of oil. Add the spaghetti and cook, stirring occasionally. Drain the spaghetti and return to the pot. Combine the onion and tomato mixture in the pot with the spaghetti, add ½ cup/120 ml/4 fl oz of the broth, along with the clams and the peas. Mix well. Cover and continue cooking until the clams have opened (about 10 minutes).
- To serve, add the remaining broth (the mixture should be a little soupy) and season with salt and pepper – it should be well seasoned. Serve in small individual casserole dishes. Although best prepared at the last minute, this dish can be made in advance and reheated.

Mussels with Herbed Garlic Butter

SERVES 4

INGREDIENTS

1 kg/2¼ lb mussels

85 g/3 oz butter

2 cloves garlic, peeled and crushed

2 tablespoons finely chopped basil

½ bunch finely chopped parsley

steamed rice or noodles

METHOD

- Remove the fibrous beards from the mussels by pulling firmly. Wash the mussels well, scrubbing with a small brush if necessary.
- Combine the butter, garlic, basil and parsley. This can be done in a food processor.
- Place the mussels in a heavy-based frying pan over a medium heat.
- As the mussels start to open, spoon the herb butter over and into the mussels, continuing to add butter until all has been used. Serve on individual plates over steamed rice or noodles, which take on the delicious juices.

Spaghetti Vongole

SERVES 3–4

INGREDIENTS

285 g/10 oz spaghetti
45 ml/1½ fl oz virgin olive oil
1 onion, very finely chopped
2 cloves garlic, finely chopped
510 g/18 oz clams, cleaned and sand removed
100 ml/3⅓ fl oz white wine
salt and pepper
1 tablespoon fresh chopped oregano

METHOD

- Pre-cook the spaghetti in boiling water. Refresh in cold water, stir with half the oil and set aside.
- Heat the remaining oil in a large cooking pot over high heat. Add the onion, garlic and cook for 1 minute
- Add the clams, white wine, salt and pepper.
- When all the clams have opened, add the spaghetti and oregano. Cook for another 2 minutes and serve.

Clams in Spicy Tomato Sauce

SERVES 4–6

INGREDIENTS

1 onion
2 cloves garlic, minced
2 tablespoons olive oil
2 tomatoes, peeled and chopped
1 tablespoon tomato paste
1 teaspoon paprika
½ cup/125 ml/4 fl oz dry white wine
salt
freshly ground pepper
1 tablespoon chopped parsley
½ dried red chilli, seeds removed, crumbled
18 very small clams

METHOD

- In a shallow casserole or cooking pot, sauté the onion and garlic in the oil until the onion is wilted. Add the tomatoes, tomato paste, paprika, wine, salt, pepper, parsley and chilli. Cover and cook for 10 minutes. Add the clams, cover tightly and cook over a high heat until the clams open. Serve in the same dish.

Baked Stuffed Clams

SERVES 4–6

INGREDIENTS

18 clams
1 tablespoon olive oil
2 tablespoons minced onion
1 clove garlic, minced
6 tablespoons breadcrumbs
2 tablespoons minced cured ham
1 tablespoon dry sherry
¼ teaspoon lemon juice
salt and pepper
¼ teaspoon paprika
1 tablespoon minced parsley
butter

METHOD

- Open the clams with a knife. Chop the meat and reserve half the shells.
- In a small frying pan, heat the oil and sauté the onion and garlic until the onion is wilted. Stir in the breadcrumbs, ham, sherry, lemon juice, a little salt, pepper, paprika and parsley. Mix in the clam meat. Stuff the reserved shells with the filling.
- Dot with butter and bake in oven at 180°C/350°F for about 10 minutes, or until slightly browned.

Spaghetti with Mussels

SERVES 4

INGREDIENTS

225 g/8 oz spaghetti
100 ml/3½ fl oz olive oil
1 onion, finely chopped
3 cloves garlic, finely chopped
4 vine ripened tomatoes, diced finely
510 g/18 oz mussels, cleaned
1 tablespoon fresh aromatic herbs, chopped
100 ml/3½ fl oz dry white wine
2 tablespoon grated Parmesan cheese
1 tablespoon chopped parsley
salt and pepper to taste

METHOD

- Pre-cook the spaghetti in water. Refresh in a cold water after cooking and toss with half the oil. Set aside.
- Heat the remaining oil over a high heat in a large pot. Add the onion, garlic and tomatoes and cook for 10 minutes.
- Add the mussels, chopped herbs, white wine and salt and pepper.
- When the mussels start to open add the spaghetti.
- Stir together until all the mussels have opened.
- Serve with the Parmesan and parsley sprinkled on top.

Mediterranean Fish Stew

SERVES 6–8

INGREDIENTS

1 kg/2¼ lb mixed fish and shellfish, such as cod, snapper or mackerel fillet, raw with shells intact prawns (shrimp) and calamari (squid) tubes
510 g/18 oz mussels, cleaned
2 tablespoons olive oil
1 onion, finely chopped
1 teaspoon fennel seeds
200 ml/7 fl oz dry white wine
400 g/14 oz can chopped tomatoes
salt and pepper to taste

Rouille

2 cloves garlic, chopped
1 small red chilli, deseeded and chopped
3 tablespoon chopped fresh cilantro (coriander)
pinch of salt
3 tablespoons mayonnaise
1 tablespoon olive oil

METHOD

- First make the rouille. Crush together the garlic, chilli and cilantro (coriander) with a pinch of salt in a pestle and mortar. Stir in the mayonnaise and oil, mix well and season to taste. Refrigerate until needed.
- Skin the fish, if necessary, and cut into 5cm/2in chunks. Shell the prawns (shrimp), then slit open the back of each one and scrape out any black vein. Rinse well. Cut the calamari (squid) into 5cm/2in rings. Shell the mussels, reserving a few with shells on to garnish.
- Heat the oil in a large heavy-based saucepan and fry the onion for 4 minutes to soften. Add the fennel seeds and fry for another minute, then add the wine, tomatoes and seasoning. Bring to the boil, then simmer, uncovered, for 5 minutes, until slightly thickened. Add the fish, calamari (squid), mussels and prawns and simmer, covered, for a further 5–6 minutes, stirring occasionally, until the prawns (shrimp) are pink and everything is cooked. Season and serve with the rouille.

Seafood Paella

SERVES 8

INGREDIENTS

1 tablespoon olive oil
2 onions, chopped
2 cloves garlic, crushed
1 tablespoon fresh thyme leaves
2 teaspoons finely grated lemon rind
4 ripe tomatoes, chopped
2½ cups/500 g/1 lb short-grain white rice pinch saffron threads soaked in 2 cups/500 ml/16 fl oz water
5 cups chicken or fish stock
285 g/10 oz fresh or frozen peas
2 red bell peppers (capsicums), chopped
1 kg/2¼ lb mussels, scrubbed and beards removed
510 g/18 oz firm white fish fillets, chopped
285 g/10 oz peeled uncooked prawns (shrimp)
225 g/8 oz scallops
3 calamari (squid) tubes, sliced
1 tablespoon chopped fresh parsley

METHOD

- Preheat the barbecue to a medium heat. Place a large paella or frying pan on the barbecue, add the oil and heat. Add the onions, garlic, thyme leaves and lemon rind and cook for 3 minutes or until the onion is soft.
- Add the tomatoes and cook, stirring, for 4 minutes. Add the rice and cook, stirring, for 4 minutes longer or until the rice is translucent. Stir in the saffron mixture and stock and bring to a simmer. Simmer, stirring occasionally, for 30 minutes or until the rice has absorbed almost all of the liquid. Stir in the peas, bell peppers (capsicums) and mussels and cook for 2 minutes. Add the fish, prawns (shrimp) and scallops and cook, stirring, for 2–3 minutes. Stir in calamari (squid) and parsley and cook, stirring, for 1–2 minutes longer or until the seafood is cooked.

Fried Mussels

SERVES 8

INGREDIENTS

2 eggs, lightly beaten
salt and pepper
2 cups/250 g/8 oz breadcrumbs
1 tablespoon dry mixed aromatic herbs
1 kg/2¼ lb black mussels cleaned, cooked mariniéres style and removed from the shell (see page 100)
3 tablespoons tartare sauce

METHOD

- Place the eggs in a bowl and season with a little salt and pepper.
- Combine the breadcrumbs and herbs in a separate bowl.
- Dip the mussels in the egg mixture, then roll in breadcrumbs.
- Deep-fry the mussels in hot oil, until golden brown.
- Drain on absorbent paper and serve immediately with tartare sauce.

Lemon-Scented Fish Pie

SERVES 4

INGREDIENTS

1 kg/2¼ lb potatoes, cut into even-sized pieces
salt and black pepper
55 g/2 oz butter
1 onion, chopped
2 sticks celery, sliced
2 tablespoons plain flour
1 cup fish stock
finely grated rind and juice of 1 large lemon
510 g/18 oz white fish, cut into cubes
170 g /6 oz cooked and shelled mussels
2 tablespoon chopped fresh parsley
4 tablespoon milk

METHOD

- Cook the potatoes in boiling salted water for 15–20 minutes, until tender, then drain.
- Meanwhile, melt half of the butter in a large saucepan, then add the onion and celery and cook for 2–3 minutes, until softened. Add the flour and cook, stirring, for 1 minute, then slowly add the fish stock and cook, stirring, until thickened. Add the lemon rind and juice and season with pepper.
- Preheat the oven to 220°C/425°F. Remove the sauce from the heat, stir in the fish, mussels and parsley, then transfer to an ovenproof dish. Mash the potatoes with the remaining butter and the milk. Season, then spread evenly over the fish with a fork. Cook in the oven for 30–40 minutes, until the sauce is bubbling and the topping is starting to brown.

Chilli-Spiked Mussels in Spaghetti

SERVES 4

INGREDIENTS

340 g/12 oz dried spaghetti
1 kg/2¼ lb fresh mussels
2 tablespoon olive oil, plus 1 tablespoon extra for drizzling
2 shallots, finely chopped
4 cloves garlic, chopped
145 ml/5 fl oz dry white wine
grated rind of 1 lemon
1 teaspoon dried chilli flakes
2 tablespoons chopped fresh parsley
black pepper, to taste

METHOD

- Cook the pasta according to the packet instructions, until tender but still firm to the bite, then drain well. Meanwhile, scrub the mussels under cold running water, pull away any beards and discard any mussels that are open or damaged.
- Place the mussels in a large heavy-based saucepan, with just the water clinging to the shells. Steam for 3–4 minutes over a high heat, shaking regularly, until the shells have opened. Discard any mussels that remain closed.
- Heat 2 tablespoons of oil in a large saucepan and gently fry the shallots and garlic for 5 minutes or until softened. Add the wine and boil rapidly for 5–6 minutes, until the liquid has reduced by half. Add the mussels, lemon rind and chilli and heat for 2–3 minutes. Add the pasta to the mussels, then stir in the parsley and black pepper. Gently toss over heat and drizzle over remaining oil.

Belgian-Style Mussels

SERVES 4

INGREDIENTS

2 kg/4½ lb mussels in their shells
30 g/1 oz butter
1 tablespoon vegetable oil
4 shallots or 1 onion, chopped
2 sticks celery, chopped, plus any leaves
145 ml/5 fl oz dry white wine
black pepper
145 ml/5 fl oz thickened or double cream
4 tablespoon chopped fresh flat-leaf Italian parsley

METHOD

- Scrub the mussels under cold running water, then pull away any beards and discard any mussels that are open or damaged. Heat the butter and oil in a very large saucepan, then add the shallots or onion and celery and cook for 2–3 minutes, until the shallots are translucent.
- Stir in the wine and plenty of pepper and bring to the boil. Add the mussels, cover and cook over a high heat, shaking the pan occasionally, for 4–5 minutes, until the mussels have opened. Remove from the pan and keep warm in a bowl, discarding any that remain closed.
- Roughly chop the celery leaves, reserving a few for garnish. Add the chopped leaves, cream and parsley to the cooking juices and season again if necessary. Bring to the boil, then spoon over the mussels. Garnish with celery leaves.

Marinated Mussels

SERVES 6–8

INGREDIENTS

½ cup olive oil
3 tablespoons red wine vinegar
1 teaspoon small capers
1 tablespoon minced onion
1 tablespoon minced pimiento (Spanish red peppers, homemade or imported)
1 tablespoon chopped parsley
salt and pepper
2 dozen medium mussels
1 cup water
1 slice lemon

METHOD

- Mix the oil, vinegar, capers, onion, pimiento, parsley, salt and pepper in a bowl and set aside. Scrub the mussels well, removing the beards. Discard any that do not close tightly.
- Place the water in a saucepan with the lemon slice. Add the mussels and bring to the boil. Remove the mussels as they open then cool.
- Remove the mussel meat from the shells and add to the bowl with the marinade.
- Cover and refrigerate overnight. Reserve half the mussel shells, clean them well, and place them in a plastic bag in the refrigerator. Before serving, replace the mussels in their shells and spoon a small amount of the marinade over each.

Mussels in Chervil Sauce

SERVES 2

INGREDIENTS

2 tablespoons olive oil
1 small onion, finely chopped
2 cloves garlic, minced
1 teaspoon plain flour
½ cup dry white wine
1 bay leaf
2 tablespoons fresh lemon juice
freshly ground pepper
salt
2 dozen medium mussels
1 tablespoon minced parsley

METHOD

- Heat the oil in a shallow casserole, preferably Spanish earthenware. Sauté the onion and garlic until the onion is wilted. Stir in the flour and cook for 1 minute. Add the wine, bay leaf, lemon juice, pepper and a little salt. Simmer, covered, for 5 minutes. (May be prepared ahead.)
- Add the mussels to the sauce, cover and cook until the mussels have opened. Sprinkle with the parsley and serve.

Fettuccine with Mussels

SERVES 4

INGREDIENTS

4 lb/2 kg mussels
½ cup extra virgin olive oil
12 oz/340 g fettuccini
3½ oz/100 ml dry white wine
2 oz/60 g chopped fresh parsley
2 cloves garlic, chopped
1 teaspoon crushed dried chillies

METHOD

- Scrub the mussels under cold running water, pull away any beards and discard any mussels that are open or damaged. Place 2 tablespoons of the oil in a large, heavy-based skillet, then add the mussels. Cook, covered, shaking the skillet frequently for 2–4 minutes or until the mussels open. Discard any mussels that do not open.
- Reserve 12 mussels in their shells for garnishing. Detach the remaining mussels from their shells and set aside. Discard the shells.
- Cook the pasta in plenty of boiling salted water until tender but still firm to the bite, then drain. Meanwhile, place the remaining oil, the wine, parsley, garlic and chilli in a large, heavy-based skillet and bring to the boil. Cook for 2 minutes to boil off the alcohol. Stir the mussels and pasta into the oil and chilli mixture and toss for 30 seconds to heat through. Serve garnished with the reserved mussels.

- Note: this classic dish from Naples is a wonderful mix of fresh mussels, wine, garlic and parsley. You can use clams instead of mussels if you like; they are equally good.

Special Occasions

Bouillabaisse

SERVES 10–12

INGREDIENTS

3 kg/6½ lb fish heads and bones
4 tablespoons olive oil
3 cups dry white wine
4 carrots, peeled and sliced
2 leeks, washed and sliced
2 onions, peeled and sliced
3 sticks celery, sliced
6 tomatoes, chopped
1 teaspoon peppercorns
1 bunch of thyme, tied together
1 bunch parsley, tied together
1 bunch of dill, tied together
4 fresh bay leaves
12 cups of water
salt and pepper

Soup

2 tablespoons olive oil
2 large leeks, washed and sliced
1 large fennel bulb, finely sliced
6 shallots, peeled and sliced
3 medium potatoes, peeled and diced

METHOD

- Rinse the fish heads and bones and set aside. Heat the olive oil in a deep saucepan and add the fish heads and bones. Cook the fish pieces over a high heat, stirring constantly, until the fish pieces begin to break down, scraping up anything that sticks to the bottom of the pan, (for about 20 minutes). Add the wine and simmer, stirring well. Add the prepared vegetables, herbs, bay leaves and water and simmer for 30 minutes, skimming any scum off the surface as it appears. After 30 minutes, strain the stock thoroughly, pressing on the solids to extract as much liquid as possible. Return to the heat for a further 20 minutes then add salt and pepper to taste. Set aside.
- To make the soup, heat the olive oil in a saucepan and add the sliced leeks, fennel, shallots, potatoes and saffron and cook over medium heat until all the vegetables are golden and soft, (about 20 minutes). Add the squashed canned tomatoes and reserved fish stock and bring the soup to the boil. Add salt and pepper to taste.
- Add the fish, prawns (shrimp) and mussels and simmer for 10 minutes. Add the calamari (squid) and parsley and stir gently. Remove the soup from the heat and cover. Allow to rest for 10 minutes. Meanwhile, brush the sliced sourdough bread with olive oil and grill until golden on both sides. Rub a clove of garlic over each golden slice. To serve, place a slice of grilled bread on the bottom of each soup bowl and ladle the hot soup over, making sure that everyone gets some mussels, prawns (shrimp) and calamari (squid). Add a

RECIPE CONTINUES

large pinch of saffron threads
2 x 400 g/14oz cans Italian-style tomatoes
2 kg/4½ lb assorted fish fillets, diced
570 g/20 oz large prawns (shrimp), peeled
1 kg/2¼ lb mussels, scrubbed and rinsed
510 g/18 oz small calamari (squid), cleaned
1 bunch parsley, chopped
1 loaf sourdough bread
salt and pepper, to taste

For the Rouille

2 large bell peppers (capsicums)
1 cup/125 g/4 oz breadcrumbs
3 cloves garlic
1 teaspoon red wine vinegar
1 cup liquid from soup
2 small red chillies
olive oil
salt and pepper

spoonful of rouille if desired.

- To make the rouille, roast and then skin the bell peppers (capsicums) under a hot grill. Then place bell peppers (capsicums), breadcrumbs, garlic, vinegar, soup liquid and chillies in a food processor and process. Be careful not to over-process. When the ingredients are well mixed add enough olive oil and salt and pepper to make a flavorsome paste.

Oysters and Mussels in Shells

SERVES 6

INGREDIENTS

510 g/18 oz mussels, scrubbed and beards removed
24 oysters in half shells
55 g/2 oz butter, softened
1 tablespoon chopped fresh parsley
2 tablespoons lemon juice
1 tablespoon orange juice
1 tablespoon white wine

METHOD

- Preheat a barbecue to a high heat. Place the mussels and oysters on the barbecue grill and cook for 3–5 minutes or until the mussel shells open and the oysters are warm. Discard any mussels that do not open after 5 minutes of cooking.
- Place the butter, parsley, lemon juice, orange juice and wine in a heavy-based saucepan. Place on the barbecue and cook, stirring, for 2 minutes or until the mixture is bubbling. Place the mussels and oysters on a serving platter,

Clams in Romesco Sauce

SERVES 4–6

INGREDIENTS

2 dozen very small clams
2 pimientos, homemade or imported, cut into strips
½ cup red wine vinegar
1 bay leaf
3 tablespoons olive oil
2 slices French-style bread
3 cloves garlic, peeled
12 blanched almonds
½ cup dry white wine
1 dried red chilli, seeds removed and chopped
salt
freshly ground pepper
1 teaspoon grappa (optional)
1 tablespoon chopped parsley

Fish Broth

1 small whole fish, such as whiting, head on, cleaned
¼ cup dry white wine
1½ cups water
½ bay leaf
¼ teaspoon thyme
1 small onion
1 small carrot, peeled and cut in half
6 peppercorns
salt

METHOD

- Scrub the clams well and soak overnight in water, salt, and cornmeal to rid the clams of any sand. Soak the pimientos in a bowl with the vinegar and bay leaf for 3–4 hours.
- To make the broth, place all broth ingredients in a saucepan and bring to the boil. Cover and simmer for 1 hour. Strain and reserve 1 cup. Drain the pimientos, and dry them on paper towels. Heat the oil in a large, shallow casserole. Sauté the pimientos about 2 minutes. Transfer them to a blender or processor, leaving the remaining oil in the pan. Fry the bread slices and garlic in the remaining oil, until both are golden. Add them to the blender, along with the almonds. Blend until a past forms. With the motor running, add ¼ cup/60 ml/2 fl oz of the fish broth. When it is well blended, add the remaining broth and the wine. Beat until smooth.
- Heat the saucepan again. Strain the contents of the blender into the pan and add the chilli, salt, and pepper. Arrange the clams in the pan, cover and cook over a medium flame, removing the clams as they open. Correct the seasoning and remove the saucepan from the heat. Stir in the grappa (if desired) and return the clams to the pan, making sure that the shell section with the clam meat is covered by the sauce. Cover and set aside for 1–2 hours. Reheat and serve, sprinkled with the parsley.

Crispy Baked Mussels

MAKES 24

INGREDIENTS

24 mussels, cooked mariniéres style, top shell removed (see page 100)
115 g/4 oz bacon, partially cooked, chopped
⅓ cup/80 g/2⅔ oz butter
2 spring onions (scallions), chopped
½ cup/30g/1oz soft breadcrumbs
2 tablespoons freshly squeezed lemon juice
salt and pepper
chopped chives, for garnish

METHOD

- Line a baking dish large enough to hold the mussels in one layer with crumpled foil. Add the mussels. The foil will keep the mussels upright. Scatter the bacon over the mussels.
- Melt the butter in a small frying pan, add the spring onion (scallions). Sauté until soft. Add the breadcrumbs and lemon juice and cook until the breadcrumbs are starting to crisp. Season with salt and pepper.
- Spoon the breadcrumb mixture over mussels. Bake in a 200°C/400°F oven until the mussels are heated through and the tops are crisp. Serve hot garnished with chives.

Mussel Shooters

SERVES 6

INGREDIENTS

18 Mussels, cooked mariniéres style, taken out of the shell (see page 100)

Bloody Mary Mix

200 ml/7 fl oz tomato juice
60 ml/2 fl oz vodka
a few drops of Worcestershire sauce
a few drops of Tabasco sauce
celery salt and pepper

METHOD

- Combine all the Bloody Mary ingredients and stir until combined. Refrigerate for 2–3 hours.
- Take 6 shooter glasses and add 3 mussels to each glass. Top the glasses with Bloody Mary mix and serve immediately.

Moules Mernier

SERVES 4

INGREDIENTS

2 kg/4½ lb fresh mussels
2 onions, chopped
3 shallots, chopped
2 tablespoon chopped fresh parsley, plus extra to garnish
145 ml/5 fl oz water
145 ml/5 fl oz white wine or fish stock
4 tablespoon double cream
salt and black pepper

METHOD

- Scrub the mussels under cold running water, then pull away any beards and discard any mussels that are open or damaged.
- Place the onions, shallots, parsley, water, and the wine or stock in a large heavy-based saucepan. Cook gently for 10 minutes or until the onions and shallots have softened. Add the mussels, then cover and cook for 5 minutes or until the shells have opened, shaking the pan from time to time. Place a colander over a bowl and strain the mussels. Reserve the cooking liquid and discard any mussels that remain closed.
- Pour the reserved liquid into the pan and boil for 5 minutes or until reduced by half. Remove from the heat and stir in the cream. Season, if necessary. Return to the heat and warm through, but do not boil. Divide the mussels among 4 large bowls, pour over the sauce and sprinkle with parsley

Mussel Salsa

SERVES 4

INGREDIENTS

1½ cups/375 g/13 oz marinated green shelled mussels, drained
¼ cup lime juice
1 teaspoon garlic, crushed
½ red onion, diced
1 ripe tomato, diced
small bunch cilantro (coriander) leaves, chopped
6–10 drops Tabasco sauce, depending on taste
salt and freshly ground black pepper, to taste
1 packet crackers

METHOD

- In a medium-sized bowl, combine all of the ingredients and mix well.
- Serve with crackers.

Escargot Mussels

SERVES 4

INGREDIENTS

1 kg/2¼ lb mussels, cooked mariniéres style (see page 100)

Garlic Butter

510 g/18 oz softened butter
2 cloves garlic, minced
1 tablespoon chopped fresh parsely
30 ml/1 fl oz brandy
salt and pepper

METHOD

- Remove the extra half shell and keep the mussel in one shell.
- To make garlic butter, combine all the ingredients in a bowl and mix well.
- Top up the half shell mussel with garlic butter.
- Grill the mussels until sizzling and serve with bread.

Mussels in White Wine with Garlic, Onions and Tomatoes

SERVES 4

INGREDIENTS

1 hard-boiled egg
2 tablespoons olive oil
1 small onion, finely chopped
1 clove garlic, crushed
¼ cup/15 g/½ oz fresh breadcrumbs
200 g/7 oz can peeled tomato pieces, finely chopped
24 mussels, clams or pippies, bearded and scrubbed
1½ cups dry white wine
1 bay leaf
salt and pepper
2 tablespoons finely chopped parsley
1 lemon

METHOD

- Sieve the hard-boiled egg yolk and finely chop the white.
- Heat the oil in a small frying pan, add the onion and garlic and sauté until the onion is soft but not brown (about 5 minutes).
- Stir in the breadcrumbs, tomatoes and sieved egg yolk. Cook while stirring until most of the liquid has evaporated and the mixture is a purée. Reserve.
- Place the mussels in a large saucepan, add the wine and bay leaf and bring to the boil.
- Cover, reduce heat to very low and simmer for 8–10 minutes. Remove the mussels with a slotted spoon to a warm serving dish. Discard any unopened mussels.
- Strain the mussel liquid into the tomato mixture. Stir and bring to the boil. Add salt and pepper to taste. Pour the sauce over the mussels, sprinkle with the chopped egg white and parsley. Cut the lemon into wedges or slices and garnish. Serve with crusty bread.

Béchamel-Coated Mussels with Cured Ham

SERVES 8

INGREDIENTS

18 fresh medium-sized mussels
¼ cup water
¼ cup dry white wine
1 bay leaf
85g/3oz cured ham, in very thin slices
½ cup/30g/1oz fresh breadcrumbs
1 tablespoon grated cheese, such as Manchego or Parmesan
2 eggs, lightly beaten
oil for frying

White Sauce

5 tablespoons sweet butter
6 tablespoons plain flour
¾ cup milk
salt
freshly ground pepper
dash of nutmeg

METHOD

- Place the mussels in a saucepan with the water, wine and bay leaf. Bring to the boil, reduce the heat to medium, cover and cook, removing the mussels as they open. Transfer the cooking liquid to a bowl, remove the mussel shells and place the mussels in the liquid until ready to use. Drain the liquid from the mussels, reserving ¾ cup (if there is less, add a little water).
- To make the white sauce, melt the butter in a saucepan. Add the flour and cook, stirring, for a minute or two. Stir in the reserved mussel broth, the milk, salt, pepper and nutmeg, and cook stirring constantly, until the sauce reaches boiling point. Turn off the heat and stir the sauce occasionally until ready to use. Dry the mussels well on paper towels. Wrap a piece of ham, of about the width of a mussel, around each mussel. Coat with the white sauce and place on a dish. Refrigerate for at least 1 hour, or until the sauce becomes firm. (May be prepared ahead.)
- Combine the breadcrumbs with the grated cheese. Coat the mussels with the beaten egg, then cover with breadcrumbs. Refrigerate. In a saucepan heat the oil at least 1cm/½in deep to about 195°C/380°F and fry the mussels quickly until golden. Or, better, use a deep fryer. Drain.

Mussel Crêpes

SERVES 6

INGREDIENTS

2 kg/4½ lb mussels
½ cup dry white wine
2 tablespoons chopped onion
4 parsley stalks, bruised
6 black peppercorns, crushed

Crêpes

100 g/3½ oz plain flour
2 large eggs
mussel broth (see method)
4–6 tablespoons thickened cream
4 tablespoons butter
6 tablespoons fresh parsley, chopped

METHOD

- Wash the mussels, discarding any that are open (and do not close when touched). Pull off the beards. Place the wine, onion, parsley stalks and peppercorns in a large pan and bring to a simmer.
- Add the mussels (in 2 batches) and cover. Cook over high heat for 3–4 minutes shaking occasionally, until they are open.
- Discard the shells and any that remain shut. Strain the liquid into a measuring jug and leave to cool. Taste for seasoning. Make the crêpe batter. Place the flour in a bowl or blender and work in the eggs, mussel liquid and 2 tablespoons of cream. (Don't over beat in a blender.) Allow to stand for 1 hour.
- Melt 1 tablespoon of butter in a frying pan, swirling it around. Add to the batter and stir thoroughly. To cook crêpes, heat another tablespoon of butter and swirl. Use about 175 ml/6 fl oz crêpe batter per crêpe. It is easiest to pour from a cup.
- Lift the pan and pour the batter fast into the middle of the pan and in a circle around, tilting the pan to cover the base. (If you overdo the liquid, spoon off anything that doesn't set at once. Crêpes should be thin.
- Return the pan to the heat, shaking it to make sure the crêpe does not stick. Cook for 1 minute until golden underneath, then flip over using a spatula. Briefly cook the other side. Roll and keep warm on a plate while you make more.
- Warm the remaining cream in a saucepan with the mussel bodies. Spoon the mussels and a little cream onto one edge of a crêpe, sprinkle with parsley and roll up. Cook immediately

Stuffed Mussels

SERVES 4–5

INGREDIENTS

18 medium mussels
¾ cup water
1 slice lemon
1 tablespoon olive oil
4 tablespoons minced onion
2 tablespoons minced cured ham
1 clove garlic, minced
1 teaspoon tomato sauce
1 tablespoon minced parsley
salt and freshly ground pepper
1 cup/125 g/4 oz breadcrumbs
1 tablespoon grated cheese
2 eggs, lightly beaten with 1 teaspoon water
oil, for frying

White Sauce

3 tablespoons butter
4 tablespoons plain flour
½ cup milk
salt and freshly ground pepper

METHOD

- Scrub the mussels well and remove the beards.
- Place them in a pan with the water and lemon slice. Bring to the boil and remove the mussels as they open. Do not overcook. Reserve ½ cup/120 ml/4fl oz of the mussel broth.
- Mince the mussel meat. Separate the shells and discard half of them. Heat the olive oil in a small frying pan. Add the onion and sauté until it is wilted. Add the ham and garlic and sauté for 1 minute more. Stir in the tomato sauce, the minced mussel meat, parsley, salt and pepper. Cook for 5 minutes. Half-fill the mussel shells with this mixture.
- To make the white sauce, melt the butter in a saucepan over moderate heat. Add the flour and stir for 2 minutes. Gradually pour in the reserved mussel broth and the milk. Cook, stirring constantly, until the sauce is smooth and thick. Season with salt and pepper. Remove the pan from the heat and cool slightly, stirring occasionally.
- Using a teaspoon, cover the filled mussel shells with the white sauce, sealing the edges by smoothing with the cupped side of a spoon. Refrigerate for 1 hour or more, until the sauce hardens. (May be prepared ahead.)
- Mix together the breadcrumbs and cheese. Dip the mussels into the beaten egg, then into the crumb mixture. Deep-fry the mussels filled-side down until they are well browned. Serve warm.

Sauteed Mussels

SERVES 2

INGREDIENTS

18 medium-sized mussels
½ cup water
1 slice lemon
6 tablespoons olive oil
1 small onion, minced
1 clove garlic, minced
1 tablespoon minced parsley
½ teaspoon paprika, preferably Spanish style
½ dried and crumbled, or ¼ teaspoon 1 crushed chilli

METHOD

- Place the mussels in a frying pan with the water and lemon slice. Bring to the boil and remove the mussels as they open. Do not overcook. Discard the shells and drain the mussel meat on paper towels.
- Heat the oil in a medium-sized frying pan. Stir-fry the mussels for 1 minute and remove. Add the onion and garlic and sauté slowly, covered, for about 5 minutes. Remove from the heat. Stir in the parsley, paprika and chilli. (May be prepared ahead.)
- Return the frypan to the heat and add the mussels (with their accumulated juices). Give the mussels a turn in the sauce just to heat them, remove from the heat, cover and set aside for 2 minutes before serving. The mussels may also be served at room temperature.

Mussels Ardinais

SERVES 1

INGREDIENTS

12 large mussels
15 g/½ oz butter
1 French shallot, chopped
1 large vine-ripened tomato, diced finely
thyme
salt and pepper
55 g/2 oz Emmental or vintage Cheddar cheese, grated
12 small slices of Ardenne ham or Black Forest ham

METHOD

- Cook the mussels mariniéres style (see page 100), remove from the shell and aside retaining keep half of the shells.
- Place the butter, shallots, tomato and thyme in a pot and cook over medium heat for 10 minutes. Season with salt and pepper.
- Place the mixture in the base of the half shells.
- Roll the mussels in the ham and place on top of the mixture in the mussel shells.
- Sprinkle with cheese and warm under the grill until the cheese has melted.

Mussels Florentine

SERVES 4

INGREDIENTS

1 kg/2¼ lb mussels, cooked mariniéres style and divided in half shells (see page 100)

200 g/7 oz fresh spinach, cooked and chopped, mixed with 30 g/1 oz melted butter

Mousseline Sauce

4 egg yolks

4 tablespoons water

30 g/1 oz melted butter

1 tablespoon whipped cream

salt, pepper and nutmeg

1 lemon, juiced

METHOD

- To make the mousseline sauce, whisk the egg yolks and water together over a bain marie (see glossary) until a light mayonnaise consistency.
- Add the melted butter little by little, whisking vigorously, then season and add the lemon juice.
- Add the whipped cream, and set aside (keep warm).
- Take the cooked mussels out of the shell.
- Warm the spinach in a pot and fill each mussel shell with warm spinach.
- Put the mussels back on the bed of spinach
- Place the mousseline sauce over the mussels.
- Grill until golden brown and serve.

Tuna Stuffed Mussels

SERVES 3–4

INGREDIENTS

1 small tin flaked light-meat tuna in oil (reserve the oil)
2 teaspoons red wine vinegar
4 teaspoons minced shallots
2 teaspoons minced parsley
18 large mussels
1 cup water
1 slice lemon
2 hard-boiled egg yolks
flour, for dusting
1 egg, lightly beaten
breadcrumbs
oil, for frying

METHOD

- In a cup, combine 2 tablespoons of the tuna oil, the vinegar, shallots and parsley.
- Place the mussels in a frying pan with the water and the lemon slice. Bring to a boil and remove the mussels as they open. Discard any that do not open. Remove the mussel meat and discard the shells.
- In a small bowl mash together the egg yolks and tuna. Fill each mussel with about 1 teaspoon of this mixture (the mussels have an opening into which you place the filling). Dust with flour, cover with the beaten egg, and coat with the breadcrumbs. (May be prepared ahead.)
- In a frying pan, deep-fry the mussels until golden brown. Drain on absorbent paper.
- Drizzle the oil and vinegar mixture over the mussels and serve warm.

Clams with Mushrooms and Cured Ham

SERVES 2

INGREDIENTS

3 tablespoons olive oil
115 g/4 oz mushrooms, halved or quartered
2 cloves garlic, sliced
6 tablespoons veal broth, or a mixture of chicken and beef broth
2 tablespoons diced cured ham, cut from a ½ in/3mm thick slice
1 teaspoon fresh lemon juice
½ teaspoon chilli powder
1 bay leaf
510 g/18 oz clams
1 tablespoon minced parsley

METHOD

- Heat the oil in a frying pan and sauté the mushrooms and garlic for about 2 minutes. Remove to a warm platter. Add the broth, ham, lemon juice, chili powder, bay leaf and clams. Cover and cook, removing the clams as they open. Return the clams to the casserole, making sure the clam meat is covered by the sauce. (May be prepared ahead.) Sprinkle with parsley and serve in the same dish.

Mussels in Pernod Cream

SERVES 4

INGREDIENTS

2 kg/4½ lb mussels
¾ cup dry white wine
1 tablespoon chopped parsley
a small bouquet garni (see glossary)
90 g/3 oz butter
2 medium onions, finely chopped
1 clove garlic, chopped
¾ cup hot milk
2 egg yolks
¾ cup/180 ml/6 fl oz cream
3 tablespoons Pernod or lemon juice
freshly ground pepper
2 tablespoons chopped parsley
8 slices French bread

METHOD

- Clean the mussels and place in a large pan with the wine, parsley, and bouquet garni. Cover and cook for about 5 minutes until the mussels are opened. Discard any unopened mussels. Strain the resulting broth through a fine sieve. Remove the top shell from the mussels and discard. Keep the mussels warm.
- Heat the butter in a pan and sauté the onions and garlic gently until pale golden and soft. Add the strained mussel broth and 2 cups/500 ml/16 fl oz boiling water plus the milk. Simmer for 5 minutes.
- Beat the egg yolks with the cream and Pernod or lemon juice. Stir in a little of the hot soup then return this mixture to the pan. Reheat gently, stirring all the time. Season with pepper. Place the mussels in 4 large hot soup plates, pour the hot sauce over and sprinkle with parsley.
- Meanwhile, fry the bread in the oil or butter until golden on both sides. Serve with the mussels.

Clams in Green Sauce

SERVES 1

INGREDIENTS

1 dozen very small clams

Green Sauce

2 tablespoons olive oil
2 tablespoons finely chopped onion
4 cloves garlic, minced
4 teaspoons plain flour
¼ cup dry white wine
¼ cup plus 2 tablespoons fish broth or clam juice
2 tablespoons milk
1 small bunch parsley, finely chopped
salt
freshly ground pepper

METHOD

- To make the green sauce, heat the oil in a frying pan and sauté the onion until it is wilted. Stir in the garlic. Add the flour and cook for 1 minute. Gradually pour in the wine, broth and milk and stir in the parsley, salt and pepper.
- Cook, stirring constantly, until thickened and smooth. (May be prepared ahead.) Add the clams, cover and cook over low heat, removing the clams as they open. Return the clams to the sauce, making sure the clam meat is covered by the sauce. Serve from the pan.

Clams in White Wine Sauce

SERVES 4–6

When clams (or mussels) are cooked and opened in a sauce, there is always the danger that they will release some sand. To minimize this possibility, place them for several hours or overnight, in the refrigerator in a bowl of salted water sprinkled with one tablespoon of cornmeal. The clams will release any foreign materials and at the same time will become quite plump.

INGREDIENTS

7 tablespoons olive oil
2 tablespoons minced onion
4 cloves garlic, minced
2 dozen small clams, scrubbed, at room temperature
1 tablespoon flour
1 tablespoon paprika
2 tablespoons minced parsley
1 cup semi-sweet white wine
1 bay leaf
1 dried red chilli pepper, cut into 3 pieces, seeds removed
freshly ground pepper
salt

METHOD

- Heat the oil in a large, shallow frying pan. Sauté the onion and garlic until the onion is wilted. Add the clams and cook, uncovered, over medium-high heat until they open. (If some open much sooner than others, remove them so they do not toughen. Return to the pan when all have opened.)
- Sprinkle in the flour and stir, then add the paprika, parsley, wine, bay leaf, chili, pepper, and salt, if necessary (the liquid the clams release may be salty). Continue cooking and stirring for another 5 minutes. Serve in the cooking dish if possible, and let everyone help themselves.

Mussels Riviera

SERVES 4

INGREDIENTS

1 kg/2¼ lb mussels, cooked mariniéres style, removed in half shell (see page 100)

Riviera Mix

olive oil
1 onion, finely chopped
2 garlic cloves, chopped
4 tomatoes, finely chopped
½ red capsicum (bell pepper), chopped finely
200 ml/7 fl oz white wine
2 tablespoons rosemary, thyme and basil mix, dry or freshly chopped
salt, pepper and paprika
1 tablespoon grated Parmesan cheese

METHOD

- Heat the oil, add the onion, garlic, tomatoes and capsicum (bell pepper) in a saucepan and cook slowly for 5 minutes.
- Add white wine and herbs, season with salt, pepper and paprika. Cook slowly for 30 minutes until the mixture reaches a paste consistency.
- Using a spoon, cover the mussels with the paste, top up with Parmesan. Place under the grill and heat until the Parmesan has lightly browned. Serve with foccacia bread.

Clams Waterzooi

SERVES 4

INGREDIENTS

1 tablespoon olive oil
1 kg/2¼ lb clams, cleaned
340 ml/12 fl oz dry white wine
5 cups chicken stock or fish stock
225 g/8 oz desiree potatoes, peeled and cut large dices
145 g/5 oz carrots, peeled and sliced finely
145 g/5 oz leeks (white part only), sliced and washed
145 g/5 oz celery, sliced and washed
1 tablespoon cornflour mixed with 2 tablespoons water
100 ml/3½ fl oz thickened cream
salt and pepper
1 tablespoon chopped parsley

METHOD

- Place the oil, clams and white wine in a large pot over high heat. Cook until all have opened.
- Remove the clams and set aside.
- Put the remaining broth from the pot and chicken stock together, bring to the boil, add the potatoes and cook for 10 minutes.
- Add the carrots, leeks, and celery and cook until the potatoes are cooked (around 8–10 minutes).
- Strain the cooking liquid out of the vegetables and pour it into the other pot (on high heat). Keep the vegetables and potatoes aside.
- Bring the pot to the boil, add the cornflour mixture and boil until thickened (around 1 minute). Add the cream, salt and pepper to taste and boil for another 30 seconds.
- Add the clams and vegetables, bring back to the boil and serve in a large soup bowl or in individual plates. Garnish with parsley. Serve with dry white wine.

Lemon and Garlic Steamed Mussels

SERVES 2

INGREDIENTS

2 dozen small or medium-sized mussels
3 tablespoons olive oil
2 tablespoons fresh lemon juice
2 cloves garlic, minced

METHOD

- Place the mussels in a frying pan without water. Cover and cook over medium heat, removing the mussels to a warm platter as they open. Discard any that do not open.
- Reduce the liquid in the frying pan to about 2 or 3 tablespoons, then return the mussels to the pan. Sprinkle with the oil, lemon juice and garlic and heat for 1 minute. Serve immediately, with plenty of good bread for dunking.

Baked Clams

SERVES 1

INGREDIENTS

1 dozen clams, cooked mariniéres style (see page 100)
4 tablespoons butter
2 large cloves garlic, minced
2 tablespoons minced onion
2 tablespoons minced parsley
¼ teaspoon paprika
¼ teaspoon thyme
salt and freshly ground pepper
3 teaspoons bread crumbs
¼ teaspoon olive oil

METHOD

- Remove one shell and loosen the meat from the remaining shell. Melt the butter in a frying pan and sauté the garlic, onion, parsley, paprika, thyme, salt and pepper for 2 minutes, stirring. Cover the clams with this mixture.
- In a cup, mix the breadcrumbs with the oil. Sprinkle over the clams. (May be prepared ahead.) Heat under the grill until browned.

Mussel Risotto

SERVES 4

INGREDIENTS

200 ml/7 oz olive oil
1 onion, finely chopped
2 cloves garlic, finely chopped
1 red capsicum (bell pepper), diced
285 g/10 oz arborio rice
2¼ cups dry white wine
1 kg/2¼ lb mussels, cleaned
1 tablespoon chopped fresh aromatic herbs (thyme, rosemary, marjoram)
2 tablespoons grated Parmesan cheese

METHOD

- Place the oil in a pot over medium heat.
- Add the onion, garlic, capsicum and cook for 2 minutes.
- Add rice and half the wine. Stir with a wooden spatula and cook until the rice is almost dry.
- Add the mussels and the other half of the wine.
- Add the herbs and cook until the rice and mussels are cooked. Replace the lid and stir frequently to avoid the rice sticking to the pot.
- Serve sprinkled with grated Parmesan.

Mussels with Lime Mayonnaise

SERVES 3–4

INGREDIENTS

1½ kg/3⅓ lb mussels
1 egg
2 teaspoons each of vinegar and lime juice
½ teaspoon salt
1 teaspoon French mustard
1¼ light olive oil
carrot, for garnishing
¼ bunch fresh coriander (cilantro) or chives

METHOD

- Scrub the mussels and steam them open. Remove the top shells and cut the mussels from the bottom shells, leaving them in place. Set aside, covered, in the refrigerator.
- Place the egg, vinegar, lime juice, salt and mustard in the bowl of a food processor and process for about 20 seconds. While the machine is running, add the oil in a thin, steady stream. It is important to keep the consistency of the mayonnaise thick, so stop every now and then to check. Taste for seasoning, adding some pepper if desired. The result should be a thick emulsion. To thin, add 1 or 2 tablespoons of hot water, beating well.
- Meanwhile, peel the carrot and cut into long thin strips. Arrange the carrot strips on 6 plates and top with the mussels in their half shells.
- Spoon a little lime mayonnaise on each and garnish each mussel with a little chopped coriander (cilantro) or chives.

Mussels in Garlic Butter

SERVES 2

INGREDIENTS

125 g/4 oz butter or margarine, cubed
2 cloves garlic, crushed
4 tablespoons chopped parsley
2 dozen mussels, washed, beards removed
¼ cup/15 g/½ oz chopped shallots
2 sprigs thyme
1 bay leaf
pinch salt
1 cup dry white wine

METHOD

- Cream the butter in a bowl with the garlic and parsley. Place the mussels in a saucepan with the remaining ingredients.
- Bring to the boil and simmer for 3 minutes or until the mussels open. Detach the lid of the mussels from the base and place the mussels, still sitting in the shell, on a baking sheet. Dot generously with the garlic butter and grill until the butter has melted.

Pernod Baked Mussels

SERVES 4

INGREDIENTS

285 ml/10 fl oz mussel cooking broth, strained
1 onion, chopped
1 clove garlic, chopped
55 ml/2 fl oz white wine
85 ml/3 fl oz thickened cream
30 ml/1 fl oz Pernod
juice of ½ lemon
salt and pepper
1 kg/2¼ lb mussels cleaned, cooked mariniéres style (see page 100) and divided in half shells

METHOD

- Place the broth, onion, garlic and white wine in a pot and cook over high heat, until reduced to a quarter.
- Add the cream and reduce again to half.
- Add the Pernod, lemon juice, salt and pepper.
- Cover the mussels with the sauce.
- Heat under the grill for 2 minutes and serve.

Saffron Rice with Seafood

SERVES 6

INGREDIENTS

good pinch of Spanish saffron
3 cups fish or chicken stock
2 tablespoons olive oil
2 onions, chopped
1¼ cups/275 g/9 oz long-grain rice, rinsed
225 g/8 oz green prawns (shrimp), shelled and deveined
225 g/8 oz scallops
12 mussels or baby clams, cleaned
small bunch coriander (cilantro)

METHOD

- Heat the saffron in the stock and leave to infuse. Heat the olive oil in frying pan, paella or flattish casserole dish and gently cook the onions until soft and golden. Add the rice and cook, stirring until well coated with oil. Add the hot stock, stirring until mixture comes to the boil, reduce the heat and simmer gently for 10 minutes.
- Add the prawns (shrimp) and scallops, pressing gently into the rice. Continue to cook for 5 minutes. Add the prepared mussels or clams which should open with the heat. If you have a lid, partly cover the rice. When the rice is tender (about 18–20 minutes) remove from the heat and gently fork up the rice. Sprinkle coriander (cilantro) sprigs over and serve.

Potato Gnocchi and Pesto Mussels

SERVES 3–4

INGREDIENTS

285 g/10 oz potato and semolina gnocchi (or other type)
55 ml/2 fl oz olive oil
1 onion, finely chopped
510 g/18 oz black mussels, cleaned
100 ml/3½ fl oz dry white wine
3 vine-ripened tomatoes, diced
55 g/2 oz grated Parmesan cheese

Pesto

1 handful of basil leaves
1 handful of flat leaf (Italian) parsley leaves
145 ml/5 fl oz virgin olive oil
55 g/2 oz grated Parmesan
6 cloves garlic, crushed
salt and pepper
1 tablespoon pine nuts

METHOD

- To prepare the pesto, blend all the ingredients together until you reach a smooth paste consistency.
- Cook the gnocchi in boiling water for around 5 minutes.
- While gnocchi are cooking, place the oil in a large pot on medium heat and add the onion. Cook for 1 minute.
- Add the mussels and white wine and cook with the lid on. Add the tomatoes.
- When the mussels are cooked (about 5 minutes), add the gnocchi, tomato and pesto. Mix well.
- Serve hot with grated Parmesan.

Saffron and Surf Clam Rissotto

SERVES 4

INGREDIENTS

115 ml/4 fl oz virgin olive oil
1 onion, finely sliced
4 cloves garlic, finely chopped
400 g/14 oz arborio rice
370 ml/13 fl oz dry white wine
pinch of saffron (powder or thread)
2 cups chicken stock
salt and pepper
1 tablespoon mixed fresh herbs, chopped
510 g/18 oz clams (pipies)
510 g/18 oz surf clams

METHOD

- On medium heat put the virgin olive oil, onion and garlic in a large pot and cook for 1 minute with a lid.
- Add the rice and stir with a wooden spatula.
- Add the white wine, saffron and cook slowly until the rice starts becoming dry (around 5 minutes).
- Add the chicken stock, seasoning, mixed herbs, clams (pipies) and surf clams and cook until all the shells have opened and the rice is cooked (around 15 minutes). Keep stirring frequently to avoid the rice sticking to the pot.
- Serve with salad and/or crispy ciabatta bread.

Spinach Mornay Mussels

SERVES 3–4

INGREDIENTS

1 kg/2¼ lb mussels, cleaned
2 tablespoons butter
3 tablespoons plain flour
400 ml/14 fl oz milk
salt, pepper and nutmeg
200 g/7 oz grated Cheddar
55 g/2 oz Parmesan cheese

METHOD

- Cook the mussels mariniéres style (see page 100) and remove 1 shell from each mussel.
- Melt the butter slowly in a pot. Do not allow the butter to burn.
- Add the flour and mix until very smooth, using a wooden spatula. Remove from the heat.
- Add the milk with a whisk and return to the heat. Stir with the whisk until boiling. Reduce the heat and cook slowly for 5 minutes. Season with salt, pepper and nutmeg.
- Add the cheeses. Cook for another 5 minutes on low heat, until the cheese is completely melted.
- Top up the half-shell mussels with the mornay sauce and heat under the grill until golden brown.

Smoked Salmon and Asparagus Mussels

SERVES 3–4

INGREDIENTS

200 ml/7 fl oz mousseline sauce (see page 139)

1 kg/2¼ lb black mussels, cooked and separated in half shells

2 bunch asparagus, poached in water and cut into small 1 cm/½ in long sticks

145 g/5 oz sliced smoked salmon, shredded with a sharp knife

METHOD

- Remove the mussels from the shell and top each shell with asparagus.
- Place the mussels back in the shells and sprinkle with the shredded smoked salmon.
- Top each mussel with mousseline sauce.
- Place under the grill for 5 minutes or until the mousseline has a golden brown color.

Exotic Flavors

Blue Cheese Mussels

SERVES 3–4

INGREDIENTS

2 tablespoons olive oil
1 onion, chopped
1 stick celery, sliced
½ stick of leek, sliced
1 kg/2¼ lb mussels, cleaned
100 ml/3½ fl oz white wine
55 g/2 oz blue cheese, broken into small pieces
1 handful of fresh spinach leaves
juice of 1 lemon
30 g/1 oz butter
1 tablespoon chopped parsley

METHOD

- Place the oil, onion, celery and leek in a pot and cook for 2 minutes, stirring frequently.
- Add the mussels, white wine, blue cheese, spinach, lemon juice and cook until the mussels have opened.
- Add the butter and parsley, stir and serve.

Chinese-Style Mussels

SERVES 4

INGREDIENTS

1 tablespoon sesame oil
1 kg/2¼ lb mussels cleaned
55 ml/fl oz water
1 tablespoon cornflour or arrowroot mix with 2 tablespoons cold water
1 tablespoon fresh cilantro (coriander), chopped
3 shallots, finely chopped

Sauce

100 ml/3½ fl oz oyster sauce
1 tablespoon fresh chopped ginger
1 red chilli, sliced
1 clove garlic, chopped
1 tablespoon white vinegar
1 tablespoon soy sauce
1 pinch of Chinese Five-Spice powder

METHOD

- Mix all the sauce ingredients together.
- Place the sesame oil, mussels and water in a pot and cook until the mussels start to open.
- Add the sauce and cook until all the mussels are open.
- Add the cornflour mixture and stir until the sauce thickens. This should take around 1 minute.
- Add the cilantro (coriander) and shallots. Serve with rice or noodles.

Pacific Rim Mussels

SERVES 6

INGREDIENTS

2 tablespoons vegetable oil
1 small onion, finely chopped
2 cloves garlic, crushed
1 teaspoon grated root ginger
1 teaspoon hot curry paste
1 teaspoon ground allspice
pinch cayenne pepper
400 g/14 oz can chopped tomatoes
2 lime leaves, shredded
salt and pepper
750 ml/1½ lb large fresh mussels, scrubbed and cleaned, beards removed
1 tablespoon chopped fresh cilantro (coriander), to garnish

METHOD

- Heat the oil in the bottom of a double boiler or large saucepan. Add the onion, garlic, ginger, curry paste and spices and fry gently for 10 minutes until softened.
- Add the chopped tomatoes and shredded lime leaves, cover and simmer for 20 minutes until thickened. Season to taste with salt and pepper.
- Place the mussels either in the top of the double boiler, or in a steamer set over the saucepan. Steam the mussels over the sauce for 5 minutes. Discard any mussels which do not open.
- Carefully discard one half of each mussel shell and arrange the mussels in individual serving dishes. Spoon the sauce over and serve at once garnished with the chopped cilantro (coriander).

Curry Mussels

SERVES 2

INGREDIENTS

2 tablespoons olive oil
1 small onion, chopped
1 stick celery, sliced
1 clove garlic, chopped
2 tablespoon yellow curry paste
2 cardamom pods, crushed
1 pinch ground cumin
1 kg/2¼ lb mussels, cleaned
55 ml/2 fl oz coconut cream
1 tablespoon fresh chopped coriander (cilantro)

METHOD

- Place the oil, onion, celery, garlic, curry paste, cardamom and cumin in a pot and cook slow heat for 5 minutes, stirring frequently.
- Add the mussels and coconut cream and cook over high heat.
- Cook until all the mussels have opened, stirring every minute to the ensure mussels are cooked evenly.
- Add the cilantro (coriander), stir and serve. Add chopped chilli if you like it very spicy.

Goan Curry with Clams and Raita

SERVES 4

INGREDIENTS

Raita

½ cucumber peeled, centre removed, diced
1 tablespoon chopped fresh mint
5 tablespoons plain yoghurt
juice of 1 lemon
salt and pepper, to taste

For the Curry

30 ml/1 fl oz oil
1 onion, sliced finely
2 cloves garlic, chopped
1 tablespoon cumin powder
1 tablespoon turmeric powder
2 tablespoons mild curry powder
1 tablespoon ginger powder
2 cardamom pods, cracked
1 pinch of chilli powder
¼ stick cinnamon, cracked
510 g/18 oz clams (pipies), cleaned and free of sand
45 ml/1½ fl oz water
55 ml/2 fl oz coconut cream
1 tablespoon fresh chopped cilantro (coriander) leaves
30 ml/1 fl oz oil

METHOD

- Combine the raita ingredients and set aside. Over medium heat in a large cooking pot, add the oil, onion, garlic and all the spices. Cook for 2 minutes, gently.
- Add the shellfish and water. Cook until all the shellfish have opened. Stir frequently.
- When the shellfish are opening, add the coconut cream and cilantro (coriander) leaves.
- Serve in a large bowl with basmati rice or curried vegetables, with the raita mix on the side.

Mussels in Ginger with Pesto Crumb Topping

SERVES 4

INGREDIENTS

2 kg /4½ lb mussels
100 ml/3½ fl oz dry white wine
2 cloves garlic, crushed
4 slices Parma ham, finely chopped
85 g/3 oz fresh white breadcrumbs
2 tablespoon pesto
2 tablespoon grated fresh root ginger

METHOD

- Scrub the mussels. Soak them in cold water for 5 minutes, drain, then repeat. Remove any beards and discard any mussels that are open or damaged. Place in a large saucepan with the wine and garlic. Cover and cook over high heat for 3 minutes or until the mussels open, shaking the pan occasionally. Discard any mussels that do not open.
- Remove the mussels from the pan and reserve the cooking liquid. Discard the top shell from each mussel and arrange the mussels on the half shell on a baking sheet. Strain the mussel liquid through muslin or a clean kitchen cloth. Combine the ham, breadcrumbs, pesto and ginger and stir in 1–2 tablespoons of the mussel liquid to moisten.
- Preheat the grill to high. Spoon a little crumb mixture onto each mussel, then cook under the grill for 2 minutes or until golden and bubbling.

Laksa

SERVES 4

INGREDIENTS

45 ml/1½ fl oz peanut oil or vegetable oil
1 onion, finely chopped
3 cloves garlic, chopped
1 tablespoon laksa paste
400 ml/14 fl oz chicken stock
1 stick lemongrass, chopped
510 g/18 oz mussels, cleaned
225 ml/8 fl oz coconut cream
145 g/5 oz rice noodles
1 lime leaf, finely chopped

METHOD

- Place the oil, onions, garlic and laksa paste in a large cooking pot and cook for 3–5 minutes over medium heat.
- Add the chicken stock and lemongrass. Add mussels and cook until the mussels start to open.
- Add the coconut cream, rice noodles and lime leaf. Cook for a further 4 minutes.
- Serve when all the mussels have opened.

Mussels Parquee

SERVES 2

INGREDIENTS

24 large black mussels
1 red Spanish onion, chopped very finely
1 red chilli, chopped very finely
85 ml/3 fl oz aged red wine vinegar
55 ml red port wine
salt and pepper, to taste
lemon wedges, to serve

METHOD

- Open the mussels raw, using a small knife.
- Combine all other the ingredients, except the lemon wedges, together and set aside.
- Place the mussels on a serving plate and top with the mixture
- Refrigerate for 5 minutes and serve with lemon wedges.

Mussels Tin Tin

SERVES 2

INGREDIENTS

55 ml/2 fl oz white wine
1 red chilli, sliced
1 stalk lemongrass, crushed
1 tablespoon fresh chopped ginger
1 clove garlic, chopped
1 tablespoon peanut oil
1 kg/2¼ lb mussels, cleaned
100 ml/3½ fl oz coconut cream
1 tablespoon fresh cilantro (coriander), chopped

METHOD

- Place the white wine, chilli, lemongrass, ginger and garlic in a pot and infuse together for 15 minutes.
- In a separate pot, gently heat the oil and mussels. Add to the ingredients above.
- Add the coconut cream and cook until the mussels have opened, stirring frequently.
- Stir in the cilantro (coriander) and serve.

Spanish Marinated Mussels

SERVES 4

INGREDIENTS

1 kg/2¼ lb mussels, cooked mariniéres style (see page 100) and taken out of the shell
1 hard-boiled egg (white only), chopped finely
2 tablespoon baby capers
2 tablespoon fresh aromatic herbs, thyme, rosemary, marjoram chopped
2 vine-ripened tomatoes, finely chopped
145 ml/5 fl oz Spanish virgin olive oil
1 tablespoon Dijon mustard
30 ml/1 fl oz old sherry vinegar
1 tablespoon fresh basil, roughly chopped
salt and pepper

METHOD

- Combine all the ingredients and marinate in the refrigerator for 2 hours.
- Serve with salad or as tapas with a glass of wine.

Clams in Black Beans

SERVES 4

INGREDIENTS

1 tablespoon sesame oil
1 kg/2¼ lb clams (pipies) cleaned and sand removed
55 ml/2 fl oz water
115 ml/4 fl oz black bean sauce
1 tablespoon corn flour, mixed with 2 tablespoons water
1 tablespoon fresh cilantro (coriander), chopped
3 spring onions, finely chopped

Black Bean sauce

4 tablespoons fermented black beans (also called salted black beans)
1 tablespoon fresh chopped ginger
1 chopped red chilli
2 garlic cloves chopped
1 tablespoon white vinegar
2 tablespoons soy sauce
1 pinch Chinese Five-Spice powder
1 teaspoon sugar
2 tablespoons vegetable oil

METHOD

- Rinse the fermented black beans thoroughly and then mince (not rinsing the beans will make the sauce too salty). Mix all the sauce ingredients, set aside for 15 minutes.
- On high heat and in a large pot, put the sesame oil, clams (pipies) and water and cook until the clams (pipies) start to open.
- Add the sauce mix and cook until all the clams (pipies) have opened.
- Add the corn flour and stir until the sauce has thickened, around 1 minute on high heat.
- Add the cilantro (coriander) and spring onions.
- Serve with rice or noodles.

Mussels in Coconut Vinegar

SERVES 4

INGREDIENTS

1½ kg/3⅓ lb mussels in their shells
6 whole cilantro (coriander) sprigs, washed and roughly chopped
3 stalks fresh lemongrass, chopped, or 1½ teaspoons dried lemongrass, soaked in hot water until soft
5 cm/2 in piece fresh ginger, shredded
½ cup water
1 tablespoon vegetable oil
1 red onion, halved and sliced
2 fresh red chillies, sliced
2 tablespoons coconut vinegar
fresh cilantro (coriander) leaves

METHOD

- Place mussels, cilantro (coriander), lemongrass, ginger and water in a wok over a high heat. Cover and cook for 5 minutes or until the mussels open. Discard any mussels that do not open after 5 minutes of cooking. Remove the mussels from the wok, discard the cilantro (coriander), lemongrass and ginger. Strain the cooking liquid and reserve.
- Heat the oil in a wok over a medium heat, add the onion and chillies and stir-fry for 3 minutes or until the onion is soft. Add the mussels, reserved cooking liquid and coconut vinegar and stir-fry for 2 minutes or until the mussels are heated. Scatter with cilantro (coriander) leaves and serve.

Shellfish in Lemongrass

SERVES 4

INGREDIENTS

5 red or golden shallots, chopped
4 stalks fresh lemongrass, cut into 3 cm/1¼ in pieces, or 2 teaspoons dried lemongrass, soaked in hot water until soft
3 cloves garlic, chopped
5 cm/2 in piece fresh ginger, shredded
3 fresh red chillies, seeded and chopped
8 kaffir lime leaves, torn into pieces
750 g/1⅔ lb mussels, scrubbed and beards removed
¼ cup water
12 scallops on shells, cleaned
1 tablespoon lime juice
1 tablespoon Thai fish sauce (nam pla)
3 tablespoons fresh basil leaves

METHOD

- Place the shallots, lemongrass, garlic, ginger, chillies and lime leaves in a small bowl and mix to combine.
- Place the mussels in a wok and sprinkle over half the shallot mixture. Pour in the water, cover and cook over a high heat for 5 minutes.
- Add the scallops, remaining shallot mixture, lime juice, fish sauce (nam pla) and basil and toss to combine. Cover and cook for 4–5 minutes or until the mussels and scallops are cooked. Discard any mussels that do not open after 5 minutes.

WEIGHTS & MEASURES

Cooking is not an exact science: one does not require finely calibrated scales, pipettes and scientific equipment to cook, yet the conversion to metric measures in some countries and its interpretations must have intimidated many a good cook.

Weights are given in the recipes only for ingredients such as meats, fish, poultry and some vegetables, though a few grams/ounces one way or another will not affect the success of your dish.

Though recipes have been tested using the Australian Standard 250 ml cup, 20 ml tablespoon and 5 ml teaspoon, they will work just as well with the US and Canadian 8 fl oz cup, or the UK 300 ml cup. We have used graduated cup measures in preference to tablespoon measures so that proportions are always the same. Where tablespoon measures have been given, these are not crucial measures, so using the smaller tablespoon of the US or UK will not affect the recipe's success. At least we all agree on the teaspoon size.

For breads, cakes and pastries, the only area which might cause concern is where eggs are used, as proportions will then vary. If working with a 250 ml or 300 ml cup, use large eggs (60 g/2 oz), adding a little more liquid to the recipe for 300 ml cup measures if it seems necessary. Use the medium-sized eggs (55 g/1½ oz) with 8 fl oz cup measure. A graduated set of measuring cups and spoons is recommended, the cups in particular for measuring dry ingredients. Remember to level such ingredients to ensure their accuracy.

English Measures

All measurements are similar to Australian with two exceptions: the English cup measures 300 ml/10 fl oz, whereas the Australian cup measure 250 ml/8 fl oz. The English tablespoon (the Australian dessertspoon) measures 14.8 ml/½ fl oz against the Australian tablespoon of 20 ml/¾ fl oz.

American Measures

The American reputed pint is 16 fl oz, a quart is equal to 32 fl oz and the American gallon, 128 fl oz. The Imperial measurement is 20 fl oz to the pint, 40 fl oz a quart and 160 fl oz one gallon. The American tablespoon is equal to 14.8 ml/½ fl oz, the teaspoon is 5 ml/⅙ fl oz. The cup measure is 250 ml/8 fl oz, the same as Australia.

Dry Measures

All the measures are level, so when you have filled a cup or spoon, level it off with the edge of a knife. The scale below is the 'cook's equivalent'; it is not an exact conversion of metric to imperial measurement. To calculate the exact metric equivalent yourself, use 2.2046 lb = 1 kg or 1 lb = 0.45359 kg.

Liquid Measurements

The scale following is the 'cook's equivalent'; it is not an exact conversion of metric to imperial measurement. To calculate the exact equivalent yourself, divide millilitres by 28.349523 to obtain fluid ounce equivelant, or multiply fluid ounces by 28.349523 to obtain millilitre equivalent.

Oven Temperatures

The Celsius temperatures given here are not exact; they have been rounded off and are given as a guide only. Follow the manufacturer's temperature guide, relating it to the oven description given in the recipe. Remember gas ovens are hottest at the top, electric ovens at the bottom and convection-fan forced ovens are usually even throughout. We included Regulo numbers for gas cookers which may assist.

To convert °C to °F multiply °C by 9 and divide by 5 then add 32.

Metric Imperial	
g = grams	oz = ounces
kg = kilograms	lb = pound
15 g	½ oz
20 g	⅔ oz
30 g	1 oz
55 g	2 oz
85 g	3 oz
115 g	4 oz/¼ lb
145 g	5 oz
170 g	6 oz
200 g	7 oz
225 g	8 oz/½ lb
255 g	9 oz
285 g	10 oz
310 g	11 oz
340 g	12 oz/¾lb
370 g	13 oz
400 g	14 oz
425 g	15 oz
1,000 g	1 kg/ 35.2 oz/2.2 lb
1.5 kg	3.3 lb

Cake dish sizes	
Imperial	Metric
6 in	15 cm
7 in	18 cm
8 in	20 cm
9 in	23 cm

Oven temperatures			
	C°	F°	Regular
Very slow	120	250	1
Slow	150	300	2
Moderately slow	160	325	3
Moderate	180	350	4
Moderately hot	190–200	370–400	5–6
Hot	210–220	410–440	6–7
Very hot	230	450	8
Super hot	250–290	475–500	9–10

INDEX

Glossary

acidulated water: water with added acid, such as lemon juice or vinegar, which prevents discoloration of ingredients, particularly fruit or vegetables.
The proportion of acid to water is 1 teaspoon per 300ml.
al dente: Italian cooking term for ingredients that are cooked until tender but still firm to the bite; usually applied to pasta.
americaine: method of serving seafood - usually lobster and monkfish - in a sauce flavored with olive oil, aromatic herbs, tomatoes, white wine, fish stock, brandy and tarragon.
anglaise: cooking style for simple cooked dishes such as boiled vegetables. Assiette anglaise is a plate of cold cooked meats.
antipasto: Italian for "before the meal", it denotes an assortment of cold meats, vegetables and cheeses, often marinated, served as an hors d'oeuvre. A typical antipasto might include salami, prosciutto, marinated artichoke hearts, anchovy fillets, olives, tuna fish and Provolone cheese.
au gratin: food sprinkled with breadcrumbs, often covered with cheese sauce and browned until a crisp coating forms.
balsamic vinegar: a mild, extremely fragrant wine-based vinegar made in northern Italy. Traditionally, the vinegar is aged for at least seven years in a series of casks made of various woods.
baste: to moisten food while it is cooking by spooning or brushing on liquid or fat.
baine marie: a saucepan standing in a large pan which is filled with boiling water to keep liquids at simmering point. A double boiler will do the same job.
beat: to stir thoroughly and vigorously.
beurre manie: equal quantities of butter and flour kneaded together and added a little at a time to thicken a stew or casserole.
bird: see paupiette.
blanc: a cooking liquid made by adding flour and lemon juice to water in order to keep certain vegetables from discoloring as they cook.
blanch: to plunge into boiling water and then in some cases, into cold water. Fruits and nuts are blanched to remove skin easily.
blanquette: a white stew of lamb, veal or chicken, bound with egg yolks and cream and accompanied by onion and mushrooms.
blend: to mix thoroughly.
bonne femme: dishes cooked in the traditional French "housewife" style. Chicken and pork bonne femme are garnished with bacon, potatoes and baby onion; fish bonne femme with mushrooms in a white wine sauce.
bouquet garni: a bunch of herbs, usually consisting of sprigs of parsley, thyme, marjoram, rosemary, a bay leaf, peppercorns and cloves, tied in muslin and used to flavor stews and casseroles.
braise: to cook whole or large pieces of poultry, game, fish, meat or vegetables in a small amount of wine, stock or other liquid in a closed pot. Often the main ingredient is first browned in fat and then cooked in a low oven or very slowly on top of the stove. Braising suits tough meats and older birds and produces a mellow, rich sauce.
broil: the American term for grilling food.
brown: cook in a small amount of fat until brown.
burghul (also bulgur): a type of cracked wheat, where the kernels are steamed and dried before being crushed.
buttered: to spread with softened or melted butter.
butterfly: to slit a piece of food in half horizontally, cutting it almost through so that when opened it resembles butterfly wings. Chops, large prawns and thick fish fillets are often butterflied so that they cook more quickly.
buttermilk: a tangy, low-fat cultured milk product whose slight acidity makes it an ideal marinade base for poultry.
calzone: a semicircular pocket of pizza dough, stuffed with meat or vegetables, sealed and baked.
caramelise: to melt sugar until it is a golden brown syrup.
champignons: small mushrooms, usually canned.
chasseur: (hunter) a French cooking style in which meat and chicken dishes are cooked with mushrooms, shallots, white wine, and often tomato.
clarify: to melt butter and drain the oil off the sediment.
coat: to cover with a thin layer of flour, sugar, nuts, crumbs, poppy or sesame seeds, cinnamon sugar or a few of the ground spices.
concasser: to chop coarsely, usually tomatoes.
confit: from the French verb confire, meaning to preserve. Food that is made into a preserve by cooking very slowly and thoroughly until tender. In the case of meat, such as duck or goose, it is cooked in its own fat, and covered with it so that it does not come into contact with the air. Vegetables such as onions are good in confit.
consomme: a clear soup usually made from beef.
coulis: a thin puree, usually of fresh or cooked fruit or vegetables, which is soft enough to pour (couler means to run). A coulis may be rough-textured or very smooth.
court bouillon: the liquid in which fish, poultry or meat is cooked. It usually consists of water with bay leaf, onion, carrots and salt and freshly ground black pepper to taste. Other additives can include wine, vinegar, stock, garlic or spring onions (scallions).
couscous: cereal processed from semolina into pellets, traditionally steamed and served with meat and vegetables in the classic North African stew of the same name.
cruciferous vegetables: certain members of the mustard, cabbage and turnip families with cross-shaped flowers and strong aromas and flavors.
cream: to make soft, smooth and creamy by rubbing with back of spoon or by beating with mixer. Usually applied to fat and sugar.
croutons: small toasted or fried cubes of bread.
crudites: raw vegetables, whether cut in slices or sticks to nibble plain or with a dipping sauce, or shredded and tossed as salad with a simple dressing.
cube: to cut into small pieces with 6 equal sides.
curdle: to cause milk or sauce to separate into solid and liquid. Example, overcooked egg mixtures.

daikon radish (also called mooli): a long white Japanese radish.
dark sesame oil (also called Oriental sesame oil): dark polyunsaturated oil with a low burning point, used for seasoning. Do not replace with lighter sesame oil.
deglaze: to dissolve congealed cooking juices or glaze on the bottom of a pan by adding a liquid, then scraping and stirring vigorously whilst bringing the liquid to the boil. Juices may be used to make gravy or to add to sauce.
degrease: to skim grease from the surface of liquid. If possible the liquid should be chilled so the fat solidifies. If not, skim off most of the fat with a large metal spoon, then trail strips of paper towel on the surface of the liquid to remove any remaining globules.
devilled: a dish or sauce that is highly seasoned with a hot ingredient such as mustard, Worcestershire sauce or cayenne pepper.
dice: to cut into small cubes.
dietary fibre: a plant-cell material that is undigested or only partially digested in the human body, but which promotes healthy digestion of other food matter.
dissolve: mix a dry ingredient with liquid until absorbed.
dredge: to coat with a dry ingredient, as flour or sugar.
drizzle: to pour in a fine thread-like stream over a surface.
dust: to sprinkle or coat lightly with flour or icing sugar.
Dutch oven: a heavy casserole with a lid usually made from cast iron or pottery.
emulsion: a mixture of two liquids that are not mutually soluble - for example, oil and water.
entree: in Europe, the "entry" or hors d'oeuvre; in North America entree means the main course.
fillet: special cut of beef, lamb, pork or veal; breast of poultry and game; fish cut off the bone lengthways.
flake: to break into small pieces with a fork.
flame: to ignite warmed alcohol over food.
fold in: a gentle, careful combining of a light or delicate mixture with a heavier mixture using a metal spoon.
fricassee: a dish in which poultry, fish or vegetables are bound together with a white or veloute sauce. In Britain and the United States, the name applies to an old-fashioned dish of chicken in a creamy sauce.
galette: sweet or savoury mixture shaped as a flat round.
garnish: to decorate food, usually with something edible.
gastrique: caramelized sugar deglazed with vinegar and used in fruit-flavored savoury sauces, in such dishes as duck with orange.
glaze: a thin coating of beaten egg, syrup or aspic which is brushed over pastry, fruits or cooked meats.
gluten: a protein in flour that is developed when dough is kneaded, making it elastic.
gratin: a dish cooked in the oven or under the grill so that it develops a brown crust. Breadcrumbs or cheese may be sprinkled on top first. Shallow gratin dishes ensure a maximum area of crust.
grease: to rub or brush lightly with oil or fat.
infuse: to immerse herbs, spices or other flavorings in hot liquid to flavor it. Infusion takes from two to five minutes depending on the flavoring. The liquid should be very hot but not boiling.
jardiniere: a garnish of garden vegetables, typically carrots, pickling onions, French beans and turnips.
joint: to cut poultry, game or small animals into serving pieces by dividing at the joint.
julienne: to cut food into match-like strips.
knead: to work dough using heel of hand with a pressing motion, while stretching and folding the dough.
lights: lungs of an animal, used in various meat preparations such as pates and faggots.
line: to cover the inside of a container with paper, to protect or aid in removing mixture.
macerate: to soak food in liquid to soften.
marinade: a seasoned liquid, usually an oil and acid mixture, in which meats or other foods are soaked to soften and give more flavor.
marinara: Italian "sailor's style" cooking that does not apply to any particular combination of ingredients. Marinara tomato sauce for pasta is most familiar.
marinate: to let food stand in a marinade to season and tenderize.
mask: to cover cooked food with sauce.
melt: to heat until liquified.
mince: to grind into very small pieces.
mix: to combine ingredients by stirring.
monounsaturated fats: one of three types of fats found in foods. Are believed not to raise the level of cholesterol in the blood.
nicoise: a garnish of tomatoes, garlic and black olives; a salad with anchovy, tuna and French beans is typical.
non-reactive pan: a cooking pan whose surface does not chemically react with food. Materials used include stainless steel, enamel, glass and some alloys.
noisette: small "nut" of lamb cut from boned loin or rack that is rolled, tied and cut in neat slices. Noisette also means flavored with hazelnuts, or butter cooked to a nut brown color.
normande: a cooking style for fish, with a garnish of prawns (shrimp), mussels and mushrooms in a white wine cream sauce; for poultry and meat, a sauce with cream, Calvados and apple.
olive oil: various grades of oil extract from olives. Extra virgin olive oil has a full, fruity flavor and the lowest acidity. Virgin olive oil is slightly higher in acidity and lighter in flavor. Pure olive oil is a processed blend of olive oils and has the highest acidity and lightest taste.
panade: a mixture for binding stuffings and dumplings, notably quenelles, often of choux pastry or simply breadcrumbs. A panade may also be made of frangipane, pureed potatoes or rice.
papillote: to cook food in oiled or buttered greasepoof paper or aluminium foil. Also a decorative frill to cover bone ends of chops and poultry drumsticks.
parboil: to boil or simmer until part cooked (i.e. cooked further than when blanching).
pare: to cut away outside covering.
pate: a paste of meat or seafood used as a spread for toast or crackers.
paupiette: a thin slice of meat, poultry or fish spread with a savoury stuffing and rolled. In the United States this is also called "bird" and in Britain an "olive".
peel: to strip away outside covering.
plump: to soak in liquid or moisten thoroughly until full and round.

poach: to simmer gently in enough hot liquid to cover, using care to retain shape of food.
polyunsaturated fat: one of the three types of fats found in food. These exist in large quantities in such vegetable oils as safflower, sunflower, corn and soya bean. These fats lower the level of cholesterol in the blood.
puree: a smooth paste, usually of vegetables or fruits, made by putting foods through a sieve, food mill or liquefying in a blender or food processor.
ragout: traditionally a well-seasoned, rich stew containing meat, vegetables and wine. Nowadays, a term applied to any stewed mixture.
ramekins: small oval or round individual baking dishes.
reconstitute: to put moisture back into dehydrated foods by soaking in liquid.
reduce: to cook over a very high heat, uncovered, until the liquid is reduced by evaporation.
refresh: to cool hot food quickly, either under running water or by plunging it into iced water, to stop it cooking. Particularly for vegetables and occasionally for shellfish.
rice vinegar: mild, fragrant vinegar that is less sweet than cider vinegar and not as harsh as distilled malt vinegar. Japanese rice vinegar is milder than the Chinese variety.
roulade: a piece of meat, usually pork or veal, that is spread with stuffing, rolled and often braised or poached. A roulade may also be a sweet or savoury mixture that is baked in a Swiss roll tin or paper case, filled with a contrasting filling, and rolled.
rubbing-in: a method of incorporating fat into flour, by use of fingertips only. Also incorporates air into mixture.
safflower oil: the vegetable oil that contains the highest proportion of polyunsaturated fats.
salsa: a juice derived from the main ingredient being cooked or a sauce added to a dish to enhance its flavor. In Italy the term is often used for pasta sauces; in Mexico the name usually applies to uncooked sauces served as an accompaniment, especially to corn chips.
saturated fats: one of the three types of fats found in foods. These exist in large quantities in animal products, coconut and palm oils; they raise the level of cholesterol in the blood. As high cholesterol levels may cause heart disease, saturated fat consumption is recommended to be less than 15% of kilojoules provided by the daily diet.
sauté: to cook or brown in small amount of hot fat.
score: to mark food with cuts, notches of lines to prevent curling or to make food more attractive.
scald: to bring just to boiling point, usually for milk. Also to rinse with boiling water.
sear: to brown surface quickly over high heat in hot dish.
seasoned flour: flour with salt and pepper added.
sift: to shake a dry, powdered substance through a sieve or sifter to remove any lumps and give lightness.
simmer: to cook food gently in liquid that bubbles steadily just below boiling point so that the food cooks in even heat without breaking up.
singe: to quickly flame poultry to remove all traces of feathers after plucking.
skim: to remove a surface layer (often of impurities and scum) from a liquid with a metal spoon or small ladle.
slivered: sliced in long, thin pieces, usually refers to nuts, especially almonds.
soften: re gelatine - sprinkle over cold water and allow to gel (soften) then dissolve and liquefy.
souse: to cover food, particularly fish, in wine vinegar and spices and cook slowly; the food is cooled in the same liquid. Sousing gives food a pickled flavor.
steep: to soak in warm or cold liquid in order to soften food and draw out strong flavors or impurities.
stir-fry: to cook thin slices of meat and vegetable over a high heat in a small amount of oil, stirring constantly to even cooking in a short time. Traditionally cooked in a wok, however a heavy based frying pan may be used.
stock: a liquid containing flavors, extracts and nutrients of bones, meat, fish or vegetables.
stud: to adorn with; for example, baked ham studded with whole cloves.
sugo: an Italian sauce made from the liquid or juice extracted from fruit or meat during cooking.
sweat: to cook sliced or chopped food, usually vegetables, in a little fat and no liquid over very low heat. Foil is pressed on top so that the food steams in its own juices, usually before being added to other dishes.
timbale: a creamy mixture of vegetables or meat baked in a mould. French for "kettledrum"; also denotes a drum-shaped baking dish.
thicken: to make a thin, smooth paste by mixing together arrowroot, cornflour or flour with an equal amount of cold water; stir into hot liquid, cook, stirring until thickened.
toss: to gently mix ingredients with two forks or fork spoon.
total fat: the individual daily intake of all three fats previously described in this glossary. Nutritionists recommend that fats provide no more than 35% of the energy in the diet.
vine leaves: tender, lightly flavored leaves of the grapevine, used in ethnic cuisine as wrappers for savoury mixtures. As the leaves are usually packed in brine, they should be well rinsed before use.
whip: to beat rapidly, incorporate air and produce expansion.
zest: thin outer layer of citrus fruits containing the aromatic citrus oil. It is usually thinly pared with a vegetable peeler, or grated with a zester or grater to separate it from the bitter white pith underneath.

First published in 2017 by New Holland Publishers Pty Ltd
London • Sydney • Auckland

The Chandlery Unit 704 50 Westminster Bridge Road London SE1 7QY United Kingdom
1/66 Gibbes Street Chatswood NSW 2067 Australia
5/39 Woodside Ave Northcote, Auckland 0627 New Zealand

www.newhollandpublishers.com

A record of this book is held at the British Library and the National Library of Australia.

ISBN: 9781742579245

Group Managing Director: Fiona Schultz
Designer: Lorena Susak
Production Director: James Mills-Hicks

Printer: Toppan Leefung Printing Limited

10 9 8 7 6 5 4 3 2 1

Keep up with New Holland Publishers on Facebook
www.facebook.com/NewHollandPublishers